George Evans

A Practical Treatise on Artificial Crown and Bridge Work

George Evans

A Practical Treatise on Artificial Crown and Bridge Work

ISBN/EAN: 9783337269067

Printed in Europe, USA, Canada, Australia, Japan

Cover: Foto ©berggeist007 / pixelio.de

More available books at **www.hansebooks.com**

ON

Artificial Crown- and Bridge-Work.

BY

GEORGE EVANS.

Second Edition, Revised and Enlarged.

WITH 547 ILLUSTRATIONS.

PHILADELPHIA:
THE S. S. WHITE DENTAL MFG. CO.
1889.

PRESS OF PATTERSON & WHITE, PHILA.

To the Members

of the

First District Dental Society of the State of New York,

this book is respectfully dedicated

by

THE AUTHOR.

PREFACE TO THE SECOND EDITION.

In a field of practice so new as that of which this volume treats it is natural that changes and improvements in methods and systems should be evolved with unceasing rapidity. Although less than a year has elapsed since the publication of the first edition, the revision of the work for the introduction of new matter has become necessary to properly fulfill the requirements of a practical and comprehensive treatise.

Professional interest in the subject of crown- and bridge-work continues unabated. Judging from the space its discussion occupies in the journals and the proceedings of societies, this branch of practice is gaining in favor. Many new ideas respecting it have reached the profession through each of the channels named. From these and from other available sources careful selection of methods and appliances possessing merit has been made which by incorporation in this second edition should enhance its value to the busy practitioner.

The author entertains a grateful appreciation of the complimentary acknowledgments which have been accorded to his efforts and intentions, and hopes for an equally favorable estimate of his labors in connection with this edition.

GEORGE EVANS.

NEW YORK, July, 1889.

PREFACE TO THE FIRST EDITION.

The aim of the author in preparing this treatise is to supply an admitted want in dental literature.

Much that is new in dentistry and much more that is, by many practitioners at least, only imperfectly understood, is involved in crown- and bridge-work. Properly practiced it approaches a fine art; but misapprehension of the principles underlying it, lack of judgment in their application, and improper practice have conspired to prevent its general acceptation by the profession, and it has in consequence been only partially indorsed or even wholly condemned when a better understanding would have insured its hearty approval. Still, its possibilities are seen to be so great that at present no other branch of dentistry more engages the attention of practitioners, and in no other is a livelier interest or a greater desire for real information manifested.

The subject is here presented almost wholly from the practical stand-point, in the belief that the utility and esthetic advantages of crown- and bridge-work may be thus more readily made apparent to the practical men in dentistry. To this end the various methods described are classified in systems, and their treatment is made as concise as their importance will permit. Due credit is given where the methods and descriptions of others are used.

Acknowledgment is gratefully rendered to Drs. H. A. Parr, of New York City, E. Parmly Brown, of Flushing, N. Y., and M. H. Fletcher, of Cincinnati, for personal assistance.

GEORGE EVANS.

NEW YORK, August, 1888.

CONTENTS.

	PAGE
INTRODUCTION	13
ARTIFICIAL CROWN- AND BRIDGE-WORK	17

PART I.

PREPARATORY TREATMENT OF TEETH AND ROOTS FOR CROWN-WORK.

CHAPTER I.

THE PULPS OF TEETH—THEIR PRESERVATION OR DEVITALIZATION—PULP CAPPING	22

CHAPTER II.

DEVITALIZATION OF THE PULP	27

CHAPTER III.

PULPLESS TEETH,—THEIR TREATMENT AND DISINFECTION	31

CHAPTER IV.

CHRONIC ALVEOLAR ABSCESS	35

CHAPTER V.

SHAPING TEETH AND ROOTS FOR CROWN-WORK	39

PART II.

ARTIFICIAL CROWN-WORK.

THE PORCELAIN SYSTEM.

CHAPTER I.

PORCELAIN CROWNS	48
The Bonwill Crown	49
The How Crowns and Methods	55
The Gates Crown	62
The Foster Crown	62
The Howland Crown	63
The Logan, Brown, and New Richmond Crowns	63
Remarks on the Use of Porcelain Crowns	72

CHAPTER II.
Porcelain Crown with Gold Collar Attachment 74

CHAPTER III.
The Weston Crown 78

CHAPTER IV.
Porcelain Crowns with Rubber or Vulcanite Attachment 81

THE GOLD SYSTEM.

CHAPTER V.
Porcelain and Gold Crown without a Collar 82

CHAPTER VI.
Gold Collar Crowns 84
    The Construction and Adaptation of Collars 84

CHAPTER VII.
Gold Collar Crowns with Porcelain Fronts 89
    Incisors and Cuspids 89
    Bicuspids and Molars 92

CHAPTER VIII.
All-Gold Collar Crowns for Bicuspids and Molars constructed in Sections 95

CHAPTER IX.
The Gold Seamless Cap Crown 104
    Incisors, Cuspids, and Bicuspids, with Porcelain Fronts 104
    All-Gold Bicuspids and Molars 106

CHAPTER X.
Gold Seamless Contour Crowns 110

CHAPTER XI.
Gold Crowns with Porcelain Fronts for Teeth with Living Pulps 119
    Collar Crowns Hygienically Considered 122

CHAPTER XII.
Special Forms of Gold Crowns with Porcelain Fronts 124
    The Parr Crown 124
    The Leech Crown 125
    The Low Crown 126
    The Perry Crown 128

CHAPTER XIII.

CROWNING FRACTURED TEETH AND ROOTS—CROWNING MOLAR ROOTS DECAYED APART AT BIFURCATION—CROWNING IN CASES OF IRREGULARITY 130
    Longitudinal Fracture of the Crown and Root 130
    Fracture of the Crown with Slanting Fracture of the Root 131
    Crowning Molar Roots decayed apart at the Bifurcation 132
    Dr. Farrar's Cantilever Crown 132
    Methods of Crowning in Cases of Irregularity 133

CHAPTER XIV.

PARTIAL CROWNS 134

CHAPTER XV.

FINISHING AND POLISHING—PROCESS OF CEMENTATION 142
    Finishing and Polishing Crown-Work 142
    Insertion and Cementation 142

PART III.

BRIDGE-WORK.

CHAPTER I.

CONSTRUCTION OF BRIDGE-WORK 152

CHAPTER II.

SPECIAL PROCESSES AND APPLIANCES IN BRIDGE-WORK 163

CHAPTER III.

EXTENSION BRIDGES 170

CHAPTER IV.

DOUBLE BAR-BRIDGES 174

CHAPTER V.

EXTENSIVE APPLICATIONS OF CROWN- AND BRIDGE-WORK 177

CHAPTER VI.

REPAIR OF CROWN- OR BRIDGE-WORK 185

CHAPTER VII.

THE HYGIENIC CONDITION OF THE MOUTH AS AFFECTED BY BRIDGE-WORK 187

CHAPTER VIII.

DETACHABLE AND REMOVABLE BRIDGE-WORK 189
    Dr. Winder's Sectional Crown Method 189
    Dr. Litch's Method 192
    Dr. R. W. Starr's Methods 194
    Dr. C. M. Richmond's Method 200
    Dr. Parr's Methods 200

CHAPTER IX.

	PAGE
Removable Plate Bridges	206
Dr. Waters's Methods	218

CHAPTER X.

The Low Bridge	222

CHAPTER XI.

Dr. Knapp's Methods	227

CHAPTER XII.

Dr. Melotte's Method	232

CHAPTER XIII.

Partial Cap and Pin-Bridge Methods	236

CHAPTER XIV.

The Mandrel System	245
Detachable Bridge-Work	257

CHAPTER XV.

Porcelain Bridge-Work	262
Dr. Brown's Method	262

CHAPTER XVI.

Crown- and Bridge-Work combined with Operative Dentistry in Dental Prosthesis	272

PART IV.

MATERIALS AND PROCESSES USED IN CROWN- AND BRIDGE-WORK.

CHAPTER I.

Plate and Solders	281

CHAPTER II.

Porcelain Teeth	285

CHAPTER III.

Molds and Dies	286

CHAPTER IV.

Soldering	288

CHAPTER V.

Instruments and Appliances	290

INTRODUCTION.

Of the origin of the art of dentistry no one can speak with certainty, as its early history is shrouded in the mists of antiquity; but dental operations are recorded in very remote times.

References are made to the art in the writings of Hippocrates, in the fifth century B.C. Martial, the Latin poet, in the first century B.C., says that a Roman dentist "Cascellius is in the habit of fastening as well as extracting the teeth." To Lelius he says, "You are not ashamed to purchase teeth and hair;" and adds that "the toothless mouth of Egle was repaired with bone and ivory;" also, that "Galla, more refined, removed her artificial teeth during the night."

Horace, in the same century, cites the case of the "sorceresses Canidia and Sagana running through the city and losing the one her false hair, the other her false teeth."

Galen, the celebrated physician, in the second century A.D., also speaks of the art of dentistry as being then practiced.

These early operations were limited to the extraction of offending teeth and the replacement of those which had been lost with substitutes which were retained in position by means of narrow bands or ligatures attaching them to the adjoining natural teeth, and without the use of plates. Crude as they were, they formed the first expression of the art of dentistry, a beneficent art from the beginning, in that it sought to restore pathological or accidental defects. Confined to the simplest operations, it existed for centuries, and then was apparently lost during the Dark Ages, to reappear when the more general diffusion of knowledge ushered in the modern era of science and invention.

After its revival, dentistry, so much of it as was known, was in a measure a secret art, the practice of which even within the memory of men now living, and they not the oldest, was involved in mystery; but recent progress has lifted the veil, and dentistry, in the treatment of the teeth on correct, scientific, rational principles, has developed an art and a science which have given it honorable rank among the professions. In its two-fold evolution it has absorbed from every available source whatever would broaden its science or perfect its art. It calls to its aid anatomy, physiology, pathology, chemistry, therapeutics, metallurgy, sculpture, and mechanics, with each of which it stands in closer or more remote relation; and the practitioners of dentistry who have become the most eminent and useful have been men of broad attainments and great versatility of talent.

In the history of all progress, movements apparently of a more or less reactionary character are recorded. In the useful arts especially it is not uncommon to find a return to forms and methods formerly used but long since discarded and forgotten. So in dentistry we find methods of treatment and modes of practice once in vogue but long fallen into disuse, revived with improvements and modifications that stamp them as practically rediscoveries.

These movements are not to be regarded as retrogressive, because the modifications which accompany the reintroduction of practical ideas and inventions stamp them as real advances, and indicate clearly that the cycle of knowledge is ever widening with experience. This volume demonstrates how modern dentistry has utilized the principles of some of the simplest original operations, and by " proving all things, holding fast that which is good," has attained its present honorable position in both its scientific and artistic departments.

The history of dentistry of later years is, in brief, a recital of progress and improvement. The medical profession officially recognized it as closely allied to medicine by inviting its representatives to take part in the late International Medical Congress on the footing of professional equality.

Such is the position which dentistry has attained. Much of the progress which has made its present elevation possible must be credited to the dental profession of the United States, which has been justly termed the cradle of modern dentistry. Here the validity of the idea that scientific knowledge should form the basis of training for practice was first demonstrated by the successful establishment of dental schools; here the first journal for the interchange among dentists of thought and experience was founded; here the first association having for its object the uplifting and upholding of dentistry by the mutual helpfulness of its practitioners had its origin; here, in a word, dentistry was first divorced from mystery, here it first passed the narrow confines of a mere handicraft and earned for itself the right to be classed among the learned and liberal professions.

ARTIFICIAL CROWN- AND BRIDGE-WORK.

MODERN artificial crown- and bridge-work belongs to the department of dentistry until recently termed "mechanical," but the judgment, skill, and scientific information required place it far above ordinary mechanical dentistry, which has sunk to a low estate since the introduction of vulcanite. To such an extent has vulcanite, by reason of its cheapness and ease of manipulation, superseded other materials demanding greater knowledge and skill in their manipulation, as to retard the higher development of prosthetic dentistry, and indeed to divest it, in the hands of those who depend upon vulcanite, of the dignity which should belong to dentistry as a profession.

But modern crown- and bridge-work, properly understood and properly performed, takes high rank in dental art, and offers wide scope for versatility of talent and inventive genius. The varied and complicated cases presenting for treatment frequently suggest to the expert novel contrivances and methods of construction and application. Successful practice of crown- and bridge-work depends upon a thorough mastery of the underlying principles, and expertness in the processes involved, governed by sound judgment and perfect candor. The interests of the patient should be paramount to every other consideration, and after a careful examination he should be given an accurate statement of the applicability of the system to his case, in respect to usefulness, appearance, durability, and comfort, as compared with other processes and appliances in use.

Surgical and mechanical operations of the most delicate nature are required. Nothing, indeed, in dentistry demands finer manipulation. A practical consideration of the subject will show that a knowledge of anatomy, pathology, and therapeutics, and as well mechanical and artistic skill, are necessary to the correct treatment of cases and the proper performance of the operations indicated. Among the principal steps in an operation may be named, first, the preparatory treatment of the natural roots and teeth for the final process, involving the diagnosis of present or probable lesions and the prescription of whatever remedial or prophylactic measures may be needful; second, in crown-work, the adaptation of the artificial crowns to the cervical portion of the natural roots and the contiguous membranes, and the restoration of the articulation and the anatomical contour; and, in bridge-work, the selection of suitable teeth or roots for foundation piers or abutments, and the choice and adaptation in constructive practice of the forms which will insure the highest degree of stability and best sustain the force of occlusion, thereby avoiding abnormal positions and conditions.

The practice of crown- and bridge-work by dentists possessing the requisite attainments and governed by correct ethical principles gives results which will establish its value, remove erroneous impressions, and insure a wide professional and public indorsement of this important branch of prosthetic dentistry. Its extraordinary facilities for preserving and replacing teeth are gradually making for it the position in dental art which it merits.

PART I.

PREPARATORY TREATMENT OF TEETH AND ROOTS FOR CROWN-WORK.

PREPARATORY TREATMENT OF TEETH AND ROOTS FOR CROWN-WORK.

Preparatory treatment of teeth and roots for crown-work includes, in addition to the shaping required to fit them for the reception of the crowns, the bringing about of the healthiest possible condition in the teeth and roots and the adjacent parts, as the cure of existing lesions, the removal of calculus where necessary, and the adoption of such measures as shall prevent the recurrence of old troubles or the inception of new.

Notwithstanding all that advanced knowledge of therapeutical agents and skill in their use permit, there are many teeth and roots which cannot be rendered suitable for the successful application of crown- or bridge-work. Roots which are permeated and softened by decay, exposed or loosened from absorption of the gums and alveoli, or affected with irremediable disease of the investing membranes, should be thus classed. Cases in which abscess with necrosis has extensively impaired the walls of the alveoli are equally intractable.

Experience shows that the results in this department of dentistry depend largely upon diathesis or constitutional tendency and upon the attention given to the preservation of the health of the mouth; and these conditions should be carefully estimated in the selection of a system of treatment and the method of its application.

CHAPTER I.

THE PULPS OF TEETH—THEIR PRESERVATION OR DEVITALIZATION—PULP CAPPING.

THE preservation of the vitality of the pulps of the teeth is a matter of as much importance in connection with crown- and bridge-work as in any other class of operations, though the excision of natural crowns for the purpose of utilizing the roots as abutments for bridge-work is extensively practiced, and is defended on the theory that the vitality of the dentine is maintained by the cementum after the extirpation of the pulp.[1]

Dr. C. F. W. Bödecker, discussing the subject of the "Distribution of Living Matter in Human Dentine," says,[2]—

"1st. The dentinal canaliculi are excavations in the basis-substance of the dentine, each containing in its center a *fiber of living* matter. Besides the dentinal canaliculi, there exists an extremely delicate net-work within the basis-substance of the

---

[1] Dr. J. L. Williams says, "The life and vitality of the cementum remain intact and uninjured, and even the dentine may, and undoubtedly does, retain a certain amount of vitality, for something analogous to a healing process takes place at the ends of the broken fibrillæ next to the pulp-chamber, and by one of those wonderful provisional conditions which we so often meet with in the economy of the animal kingdom nature reverses or changes the origin of nutritive supply, and the material for maintaining the continued vitality of the dentine comes through the cementum."

He further says, " Every practicing dentist has observed that a tooth which is removed from contact with the fluids of the mouth changes color. This change in color is largely the result of the evaporation of the water from the organic portion of the tooth. Now, if the apical foramen of such a tooth be closed, and the tooth be then placed in water, or preferably, glycerin and water, in a short time it will regain nearly its original color, and at the same time it will be found that it has increased in weight. This means, of course, that the entire tooth has absorbed from the surface a certain quantity of the fluid, and this fluid has penetrated every part of the solid structure of the tooth. Will any one doubt, with these facts in view, that when the tooth is in position in the jaw, and surrounded by all the delicate adjustments furnished by nature, there may be a circulation of nutrient fluids throughout the entire root after the removal of the pulp?"

It is difficult to understand how a saturation can be compared to an infiltration controlled by vital circulation.—G. E.

[2] *Dental Cosmos*, vol. xx, page 656.

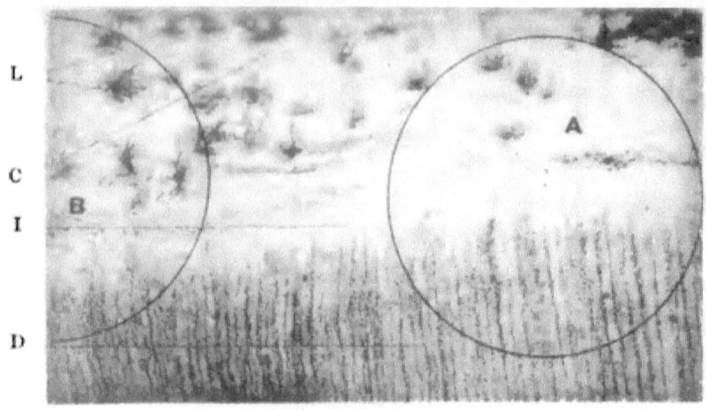

PLATE II.

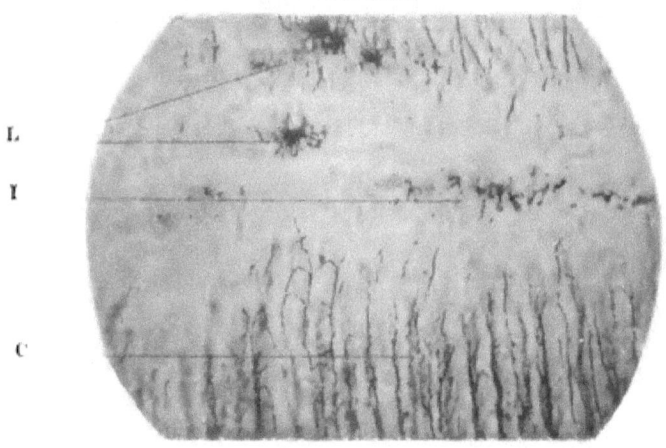

PLATE III.

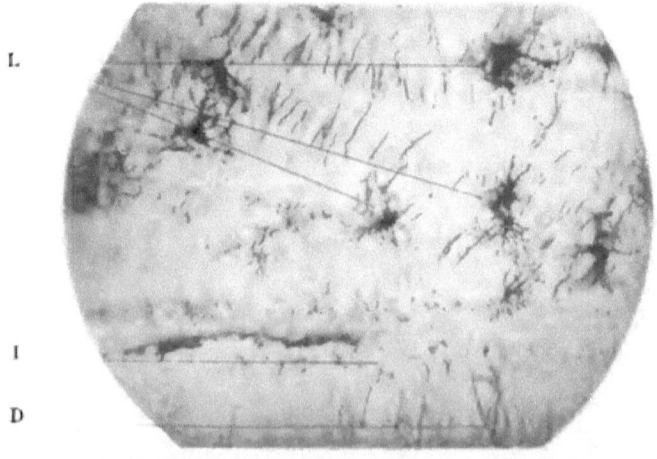

FROM SPECIMENS MADE BY M. H. FLETCHER, M. D., DENTIST.   CINCINNATI, OHIO.

dentine, into which innumerable offshoots of the dentinal fibers pass. Although we cannot trace the living matter throughout the whole net-work in the basis-substance, evidently owing to its delicacy, we are justified in assuming that not only the dentinal canaliculi, but the whole basis-substance of the dentine is also pierced by a delicate net-work of living matter. The living matter of the dentine is in direct union with that of the protoplasmic bodies of the pulp, of the cementum, and of the enamel.

"2d. The cementum, as well as ordinary bone, is provided with lacunæ and canaliculi. The lacunæ contain nucleated protoplasmic bodies, and the canaliculi hold offshoots of the living matter of the protoplasm. The whole basis-substance of the cementum is traversed by a delicate net-work, which in all probability contains living matter, though this is traceable only in its thorn-like projections from the periphery of the protoplasm and its larger offshoots. The living matter of the cementum is uninterruptedly connected with that of the periosteum, and continuous with that of the dentine, either through intervening protoplasmic bodies in the interzonal layer, or directly with the dentinal fibers."

This distribution and relative connection of living matter as described refers to an existing state of perfect vitality of all the parts. When the pulp loses its vitality, an entirely different condition results. The tubuli are then deprived of vital circulation, except along the line of the outer portion of the dentine, where, in the interzonal layer, the fibrillæ anastomose with the living matter of the cementum. The vitality supported by this anastomosis is almost entirely confined to this part, the nutrient supply being insufficient to assume the functions of the pulp and maintain circulation in the main body of the dentine. (See Plates I, II, III.)

[1] PLATE I.—Longitudinal section of the root of a superior bicuspid at junction of dentine with cementum. C, cementum; D, dentine; I, interzonal layer; L, lacunæ of cementum   175.

PLATE II.—A field taken from Plate I in position marked A. L, lacunæ of cementum; C, canaliculi of dentine; I, interzonal layer   210.

PLATE III.—A field taken from Plate I in position marked B. L, L, L, lacunæ of cementum; I, interzonal layer; D, dentinal tubes and their nearest approach to the lacunæ.   210.

A study of these plates demonstrates the limited nature of the anastomosis of the fibers of living matter of the dentine and cementum.

Chemical analysis of the dentine shows that the organic matter, consisting principally of the fibrillæ, exists in the proportion of nearly 28 per cent. to 72 per cent. of lime-salts.[1]

When the pulp has been removed, the devitalized fibrillæ still remain, and are capable of generating septic gases which will exert an influence in producing irritation of the cementum and pericementum, no matter how thoroughly the root-canal has been filled and the foramen closed. An examination of the investing membranes of pulpless teeth almost always shows the existence of abnormal conditions, by which their firmness is to some extent impaired, their susceptibility to acute inflammation increased, and their reliability as foundations for crown- or bridge-work greatly lessened when compared with teeth which have living pulps.

In crown-work, facilities are afforded for the preservation of pulps in the posterior teeth. Thus, in a case verging on exposure, only partial removal of the decay is usually necessary, as, when the operation is completed, the natural crown will be hermetically inclosed by the artificial one.

Extirpation is demanded for those pulps whose permanent preservation cannot be placed beyond doubt, as failure involves more serious consequences in crown- and bridge-work than in filling-operations. The lesions of the pulp which seem to require its extirpation, according to the generally expressed opinion on the subject, are exposure with hypertrophy, rupture of the pulp with exudation of plasma in which pulsation is visible, congestion, and pulpitis which does not yield promptly to remedial treatment.

The operation of capping a nearly or partially exposed pulp should include, as a necessary precaution against subsequent irritation, the thorough disinfection of any remaining decom-

---

[1] According to Berzelius and Bibra, dentine consists of

Organic matter (tooth-cartilage)	27.61
Fat	.40
Calcium phosphate and fluoride	66.72
Magnesium phosphate	1.18
Calcium carbonate	3.36
Other salts	.83

Age lessens the proportion of living matter and increases the percentage of lime-salts.

posed dentine. An **excellent method** of securing disinfection is by first thoroughly **washing** the cavity several times with tepid water thrown gently **from the large** point of a syringe around the sides of the cavity; then, taking measures to prevent the entrance of saliva, wiping the cavity with absorbent cotton and passing over its surface a light current of hot air from a hot-air syringe. The heat should be sufficient **to cause some discomfort to** the patient, but **not** enough to produce irritation **of the pulp.** The dried cavity is then **immediately** saturated with wood creasote[1] **previously warmed**[2] **to the normal temperature of the body by holding the pellet of cotton on which it is applied** over the flame of a lamp for a moment. The creasote relieves any pain caused by the evaporation of moisture, and disinfects and sterilizes any decomposed matter in proximity to the pulp. The object of the application of the creasote having been accomplished, it should then be removed as completely as possible. **To this end the cavity should** first be wiped with absorbent **cotton, and hot air again introduced to** evaporate the creasote **sufficiently to give a dry appearance to the** surface. This second application of hot air, owing to the **effect of the** creasote, **will cause very little or no pain.** The **pulp is then** capped with **oxyphosphate. For this purpose the cement should be used soft. The proper quantity is then applied to one side of the cavity and brought over against the bottom in such a way as not to** inclose **air between the cement and the surface, or cause the** slightest **pressure upon the pulp. This is an operation requiring** careful **and delicate manipulation.** Properly performed, **it is preferable, in most** cases, **to** protecting the **part with** a **plate or cap, of either a** metallic or non-metallic substance, fitted **to the bottom of the** cavity to avoid pressure **of the** cement; or **to the use of concave** caps filled with the **cement; as,** owing **to the adhesive** character of oxyphosphate of zinc, pressure from encompassed air is apt **to occur** frequently when the cap is adjusted in position.

---

[1] A refined pure wood creasote, such as is prepared for dental uses, is the best for **this** purpose. **Carbolic** acid, alone **or in** combination with oil of cloves, is preferably used by **some** operators.

[2] Thermal shock **to the pulp is as** unwarranted from the application of cold creasote as if produced **in any other** manner.

Where capping a pulp is necessary, it should be preliminary to any other operation to be performed. A non-vital condition of the pulp in one root of a tooth contraindicates any attempt to preserve it in any of the other roots, in connection with crown- and bridge-work. The rubber-dam, when its use is practicable, will be found a material aid in difficult pulp-capping operations.

CHAPTER II.

DEVITALIZATION OF THE PULP

In preparation for crown-work two methods of devitalization are practiced—the heroic,—instantaneous devitalization, or extirpation,—and gradual devitalization by arsenical treatment.

Instantaneous devitalization is accomplished by first administering to the patient sufficient nitrous oxide to produce partial anesthesia, then with a drill quickly opening into the pulp-chamber, and *lacerating* the pulp well up the canal with a probe or smooth broach. Instantly afterward a pellet of cotton, saturated with carbolic acid, is forced up the canal, and, if possible, left until the next day, when the pulp will be found in a coagulated mass that is easily removed entire.

Devitalization of the pulp as just described is practicable only in teeth in normal condition. In acute inflammation, after laceration of the pulp, warm water should be gently injected into the pulp-chamber, and sedative agents then applied. Subsequent treatment should be such as will complete the devitalization and extirpation of the pulp.

Excision of the crown and instantaneous extirpation of the pulp is practiced by many as follows: Two parallel grooves are cut opposite to each other, through the enamel deep into the dentine, one on the labial portion of the tooth and the other on the palatal wall, close to the gum, with a rapidly revolving rubber and corundum disk (Fig. 1). Then with excising forceps, the cutting-edges of which are inserted in these grooves, the crown is quickly severed from the root (Fig. 2). The pulp either adheres to the excised crown, leaving the canal empty, or remains in the root, fully exposed. In the latter case,

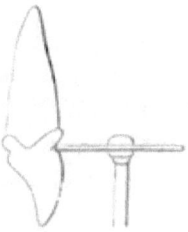

Fig. 1.

a pointed piece of orange wood, previously cut to fit the canal, and saturated with carbolic acid, is quickly driven with a blow into the pulp toward the apex of the root (Fig. 3). When the wood is withdrawn, the pulp usually adheres to it; if not, the wood is instantly reinserted, cut off and drilled out with the pulp, using Gates-Glidden drills in the upper portion of the canal. In this operation, it is claimed, only trifling pain is experienced by the patient, as the pulp is paralyzed by shock in the excision of the crown, or by being forced upward toward the foramen and against the walls of the canal.

Fig. 2.

The objections to this operation are, that if the pulp is not successfully extracted entire with the wood, the canal becomes filled with clotted blood, which is difficult to remove from the extreme end; also that the root and socket are *jarred* by the forceps in excising the crown; but, expertly performed, it is advantageous in many cases, though it must be confined to pulps in *normal* condition.

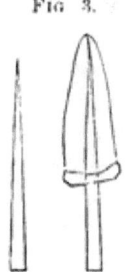

Fig. 3.

Devitalization with Arsenious Acid.—The methods of devitalizing previously described are applicable principally to the pulps of incisors and cuspids. In the posterior teeth, which do not so well permit the heroic treatment, recourse is had to arsenious acid, notwithstanding the numerous objections to its use. Two theories as to the mode of its action in devitalizing are widely entertained: First, that by producing irritation of the pulp it causes its expansion, which stops circulation by strangulation at the foramen; the other is thrombosis.[1]

This theory of thrombosis accounts for the gradual devitalization of the pulp toward the foramen, and is very likely the true explanation.

Whatever the action of arsenic on the pulp may be, it always causes an infiltration of the tubuli of the dentine with certain

---

[1] See Dr. L. C. Ingersoll's "Dental Science, Questions and Answers," page 96.

constituents of the blood, probably the liquor sanguinis. The residue of the infiltration, after the devitalization of the pulp, remains in the tubuli, and increases the difficulty of producing an aseptic condition of the dentine. It is asserted that arsenic produces devitalization of the fibrillae as far as the cementum, and, in some cases, even involves that tissue, while after instantaneous extirpation of the pulp the vitality of the dentine is to some extent preserved by the circulation it receives from the cementum.

Practical experience shows that usually instant devitalization or extirpation is the most satisfactory in general and final results. Arsenic, when used, should be applied directly to the pulp in the smallest quantity possible to effect its devitalization, and securely sealed in the cavity. The application should be kept in position no longer than is necessary to effect the devitalization of the pulp.

PLATE IV.

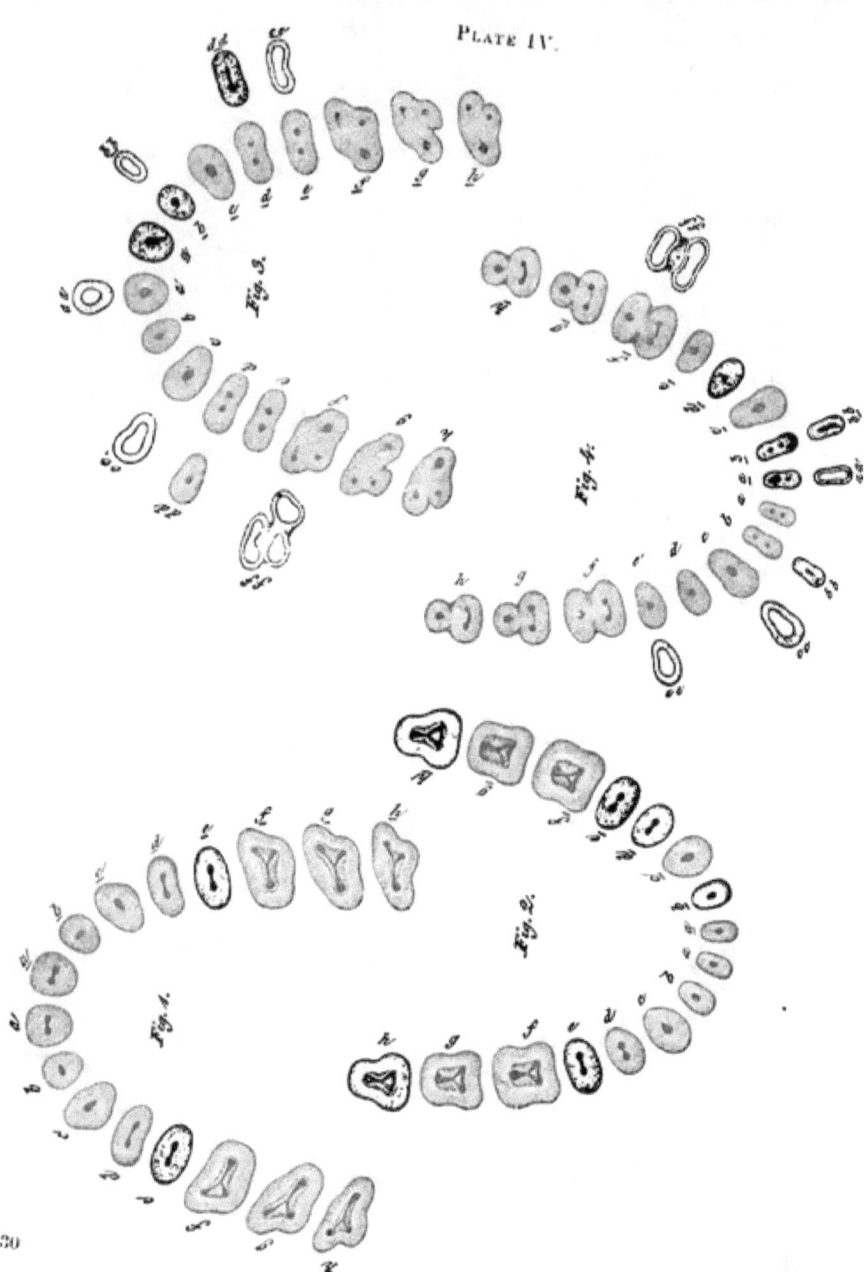

CHAPTER III.

PULPLESS TEETH,—THEIR TREATMENT AND DISINFECTION.

The treatment of pulpless teeth or roots consists in as thorough a performance as possible of the following operations:

1st. Removal of the contents of the canals.

2d. Disinfection of the root-canals and the dentine, and the establishment of permanent aseptic conditions by mummification of the contents of the tubuli.

3d. Closure of the apical foramen.

A knowledge of the usual positions of the root-canals in the different teeth is essential for a generally successful performance of these operations, which are greatly facilitated, in crown-work, by the ease with which direct access to the root-canals is obtained. (See Plate IV.)[1] An opening is first made into the center of the pulp-chamber in a line with the root-canals sufficient to give free and direct access to them, and any remaining portion of the pulp removed with broaches. The canals are then, guided by frequent explorations with a fine probe, carefully enlarged with Gates-Glidden drills (Fig. 4). At least three sizes—large, medium, and small—of drills each for the right-angle and the direct hand-piece are required. Very little, if any, pressure should be put upon them when in motion, as they

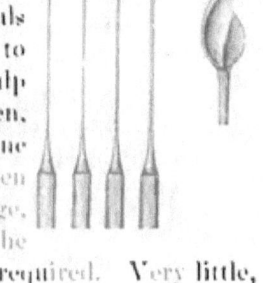

Fig. 4

---

[1] PLATE IV.—Figs. 1 and 2 represent the superior and inferior teeth in transverse section through the base of the pulp-chamber in the crown, showing the entrance to the root-canals.

Figs. 3 and 4 represent the superior and inferior teeth in transverse section through the root-canals as they diverge from the pulp-chamber.

aa, *bb*, *cc*, *dd*, *ff*, *dd*, and *ee*, Figs. 3 and 4, show the relative shapes, whether circular, oval, or flattened, of the root-canals in the teeth they severally represent.

will move forward of themselves. Under pressure a false passage in a curved root is possible, or the small drill might be broken off or forced through the apical foramen with disastrous consequences, where alveolar abscess did not exist. Neither should they be forced into canals closed by calcification. A slight pain

Fig. 5.

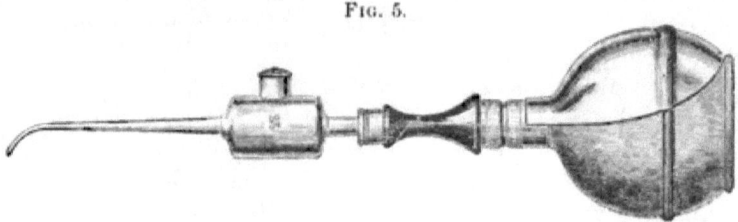

is usually experienced when they enter the zone of sensibility formed by the cementum which composes the end of the root.

Fig. 6.   The probe-like points of these drills do   Fig. 7. not cut, but simply guide the drills and confine them to the line of the canal. The occasional quick withdrawal of the drill from the canal during the process of drilling will aid removal of the *débris*. The use of these drills is condemned by some for reasons attributable to their careless or improper employment, but they are indorsed, in experienced hands, for their adaptability to the work under consideration.

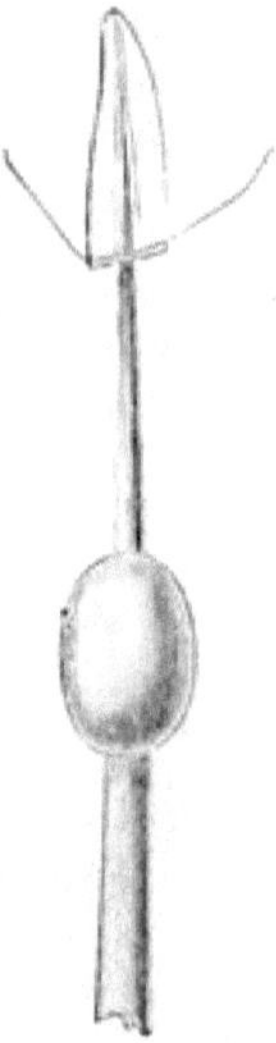

The cavity of the root-canal having been properly opened up and its contents removed, it is syringed out with tepid water, and, in case the pulp has been long devitalized, with peroxide of hydrogen, and then dried with cotton. The mouth is then properly protected with napkins, and a current of hot air introduced into the pulp-chamber by a hot-air syringe (Fig. 5), at a temperature higher, as it leaves the nozzle, than is comfortable for the finger. This heats any part of the crown remaining, and the lower portion of the root. A silver probe, tapered as fine as a broach at the point, and connected with an oval-shaped mass of silver or copper (Fig. 6) which has been heated to a dull red heat, is then introduced into the root-canal (Fig. 7). As silver possesses remarkable properties as a thermal conductor, the heat is transmitted to the point of the probe very rapidly.[1] The probe being inserted as far as possible up the canal, the patient is directed to raise the hand as a signal should the heat cause pain, when the probe must be moved up and down, or withdrawn for a moment. This procedure, following the previous application of hot air with the syringe, evaporates the moisture and gases from the root-canals and the open ends of the tubuli. The heat also acts as a germicide,[2] and carbonizes any organic matter the instrument reaches.

While in this heated and dried-out condition the dentine is saturated with an antiseptic agent, which should possess such properties as to make its action efficient and its effects permanent. At present, experience finds bichloride of mercury—$\frac{1}{500}$ solution—or iodoform most suitable for the purpose.[3] When both agents are used, the sublimate solution should be applied first, then the iodoform, after the canal has been dried. Iodoform is most conveniently applied in a saturated solution in sulphuric ether. Carbolic acid and like agents are being discarded for various reasons, among which are their inefficiency and their eventual absorption from the dentine.

After one or more treatments in the manner described, the canal is dried and the foramen closed. Of the many materials used for the purpose, gutta-percha is most approved. When used in the form of chloro-gutta-percha, a good method is to dip

---

[1] As a test of the thermal conductivity of silver, heat the oval portion of this instrument to a dull red heat, and place the broach-like point of the probe in a little water on the end of a cork. The water will boil and evaporate in a few seconds.

[2] To accomplish the same end, Dr. H. C. Register uses fine tubes of platinum, through which hot air is forced, but this is hazardous near an open foramen, less effective, and less convenient than the method detailed.

[3] Iodol or the dental tincture of iodine can be used in place of iodoform should the odor of the latter render it objectionable. A small quantity of oil of wintergreen added to the solution of iodoform aids in disguising the odor.

a few shreds of iodoformized cotton[1] in the solution, press out the surplus, and gently pack (not push) them lengthwise into the extremity of the canal. Cotton so prepared is incorruptible, and when properly inserted it effectually closes the foramen, obviates any danger of air or the solution of gutta-percha being pressed through, and is easily removed should any subsequent condition require it. The canal is then filled with gutta-percha or any other material preferred. Another method is to prepare an orange or hickory wood point to fit the canal, immerse it in the solution of gutta-percha, and then press it to position in the canal. Ordinary gutta-percha should be used to close a foramen when an abscess has just been treated by injecting through it.

Successful treatment of pulpless teeth depends not on "immediate root-filling after extirpation of the pulp," *but upon immediate root-filling after producing a thorough aseptic condition of the canals and dentine.*

Ample room should be left in any root-canal which is to receive the post of a crown, as any part of the canal not occupied by the post will be filled by the retaining material.

A pulpless tooth presented for crowning, the roots of which have been treated and filled in some previous operation, should be carefully examined, and if any doubt is entertained as to its hygienic condition it should receive the antiseptic treatment above described, as the ultimate success of crown-work depends largely upon the thoroughness of these preliminary operations.

---

[1] Cotton or wood points may be iodoformized by immersing them for a short time in a saturated solution of iodoform in ether, and then exposing them for a little while to evaporate the ether. The prepared points should be kept in a tightly-corked bottle. When this plan is followed, the odor of iodoform is almost imperceptible in the operating-room.

CHAPTER IV.

CHRONIC ALVEOLAR ABSCESS.

MANY teeth and roots presented for crown-work are affected with chronic alveolar abscess. A general description of an effective method of treatment is therefore properly associated with a discussion of the subject.

The cause of chronic alveolar abscess will be found in a continuation of those conditions which originally produced the acute form. The tooth or root being pulpless, septic gases, generated by the decomposition of organic matter in the root-canal and in the tubuli of the dentine, find an outlet through the open foramen into the apical space, causing pericementitis and formation of pus. The general treatment consists in the removal of all septic matter and gases from the root-canal and dentinal tubuli, the destruction of the pus-sac, the application of suitable therapeutic agents, and the adoption of measures to prevent further formation of pus.

Chronic alveolar abscess is usually found in the following forms: 1st. Abscess with a fistulous opening in the gum, and accessible through the root-canal and foramen of the root. 2d. Abscess with fistulous opening but not accessible through the apical foramen. 3d. Abscess from which pus discharges through the apical foramen and root-canal with no opening through the gum.

In the treatment of abscess of the first form, the canal should be enlarged as described in the treatment of pulpless teeth, and the foramen opened, if possible, with a smooth broach without the use of a drill. Tepid water is then forced through the foramen with a fine-pointed syringe (Fig. 8) introduced well up the canal, and packed in with gutta-percha, or pumped up with cotton on a broach until it passes into the abscess and out through the fistula. Peroxide of hydrogen is next used in the same manner, until it ceases to foam as it passes from the abscess. Aromatic sulphuric acid, either pure or diluted, is then applied as a germicide and powerful astringent.

In abscesses of the second form, where it is impracticable to treat through the foramen, the canal should be thoroughly disinfected, and a direct opening into the abscess effected by the track of the fistula, enlarging it if necessary. The abscess should then be thoroughly injected with peroxide of hydrogen and afterwards with aromatic sulphuric acid, by introducing the fine point of a syringe into its deepest parts. The fistula must be kept open by inserting in it, at each injection, a strand of twisted cotton saturated with oil of cloves, the patient being directed to remove it in a few hours, or the next day, for which purpose the end should be left protruding. When the apical foramen is open, one injection is usually sufficient to cure an abscess; but when the foramen is closed and the abscess is treated through the gum, several injections are sometimes necessary.

FIG. 8.

In case of "blind abscess," the third form, first clean and disinfect the root-canal, then at intervals inject the abscess through the foramen with peroxide of hydrogen until the formation of pus ceases, placing cotton saturated with oil of cloves loosely in the canal to exclude foreign substances. Should this treatment fail, an opening through the gum into the abscess must be obtained either with a lance and drill, or a trephine, and the same course pursued as in the first form of abscess.

An entrance into the apical space can be made almost painlessly in the following manner, as described by Dr. G. V. Black:[1] "The mucous membrane is first dried at the point at which it is desired to make the opening, and napkins are so placed as to

---

[1] American System of Dentistry, vol. i, page 928.

keep it dry. Then a plugging-instrument with fairly sharp serrations and of convenient shape is selected. The point of this is dipped into a 95-per-cent. solution of carbolic acid, and a drop conveyed to the mucous membrane; this will at once produce a white eschar. Then a slight scratching motion with the serrated point is begun, with the view of removing the tissue that is whitened. This is continued until the carbolic acid is thick with the débris of the tissue torn up, then it is dried out and another drop added, as before, and the process continued. This is repeated as often as may be necessary, going deeper and deeper into the tissue in the desired direction until the bone is laid bare. Then a fresh drop of the acid is placed on the bone and the periosteum carefully raised over a sufficient space; then with a sharp chisel cut through to the peridental membrane. This will generally cause some pain and some bleeding, but after giving a little time for this to cease, and adding more of the acid, the apical space can usually be reached without difficulty. No blood should be drawn at any time during the operation, except in penetrating the wall of the alveolus. In doing this no tissue is removed until it is anesthetized by the carbolic acid. This is a little tedious, but it is almost painless, and the general effect is usually better than by other modes of penetrating the apical space. The carbolic acid has the effect of modifying the pain, and the opening left does not close so readily."

After the abscess has been cured, the root-canals are treated and filled as described on page 31.

In place of aromatic sulphuric acid, if preferred or should the case suggest it, either carbolic acid, the sublimate solution ($\frac{1}{1000}$), or any other suitable therapeutic agent can be used, but most of them will be found less prompt and less effective, especially if a slightly necrosed state of the wall of the alveolus exists.[1]

Amputation of the Apex of a Root.—In long-neglected alveolar abscess, the pus-cavity occasionally involves the alveolus in such

---

[1] For an extensive consideration of this subject the reader is referred to Dr. J. N. Farrar's articles on "Sulphuric Acid v. Creasote in Treatment of Alveolar Abscess," commencing *Dental Cosmos*, vol. xx, No. 7, and Dr. G. V. Black's article in the "American System of Dentistry," vol. i, page 929.

a way as to destroy a considerable portion of the pericementum of the end of the root. The cementum of that part is consequently devitalized, and the portion of the root affected becomes degenerated in structure, and saturated with septic matter. In this condition it acquires the character of a foreign substance, proves a constant source of irritation, and defies all efforts of the membranes to perfectly inclose or encyst it.

Fig. 9.

In such cases amputation of the portion of the root which is denuded of pericementum is the best course to pursue. An opening is made in the soft tissues above the affected part with the lancet or trephine, and gradually enlarged with a tent of lint or cotton until the diseased territory is fully exposed (Fig. 9), when the devitalized end of the root and any necrosed bone in the territory are removed with a fissure-drill, and the end of the root smoothed. The root-canal is then closed with gutta-percha passed through from within, the surplus being trimmed off on the outside. Cocaine can be used in this operation.

The orifice of the cavity in the gum should be kept open until the cavity is filled by granulation. When the healing process is completed, crown-work can be proceeded with.

The amputation of roots requires skill and experience, and had better be confined to the incisors.

CHAPTER V.

SHAPING TEETH AND ROOTS FOR CROWN-WORK.

The principles governing the process of shaping a natural crown or root for any style of artificial crown with a collar attachment require that the cervical portion of the natural crown and root shall be given a form that has longitudinally-parallel sides gauged to the line of the periphery of that part, and that any of the coronal section present below it shall be reduced at least sufficiently in size to come within this line. Such a form is necessary to admit of a perfect adaptation of the collar.

The coronal section of a natural crown to be prepared is usually first ground on the occluding surface with as large a corundum-wheel as the case will conveniently admit (Fig. 10).

Fig. 10.

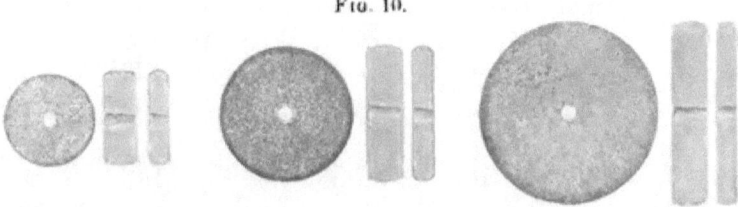

Molars and bicuspids for all-gold crowns should have enough substance removed to make a small space between them and the antagonizing teeth. The approximal surfaces are removed straight from the cervical border to the occluding surface, using diamond or rubber and corundum disks (Fig. 11) and files, and last of all, as injury to the approximal teeth is then more easily avoided, the labial and palatal portions, for which small corundum points (Fig. 12) and wheels are best adapted. The corners are then rounded. The cervical portion, which includes the junction of the dentine and enamel, is trimmed so that the sides as illustrated at A. Fig. 13, are level and parallel with the line of the root, and as deep as the collar is to be placed (Fig. 14). For this purpose, small corundum points, trimmers, and files can be used. Fig. 15 illustrates a drill, which in the ordinary hand-

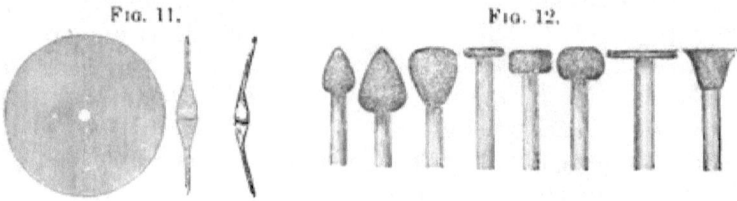

Fig. 11.   Fig. 12.

piece or in the right-angle attachment will easily and quickly accomplish this. Fig. 16 illustrates another form which can be used in a hand-socket, bracing the hand by resting the thumb on the adjoining teeth. The points should be tempered very hard. Files shaped as shown in Fig. 17 are useful in rounding angular portions. A smooth, level surface should be given the

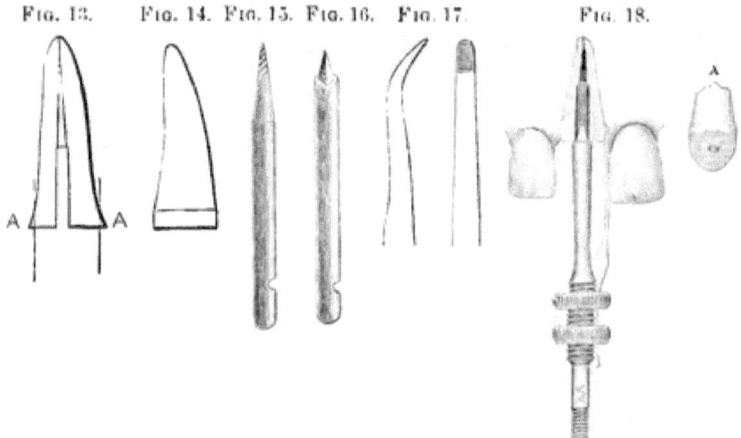

Fig. 13.   Fig. 14. Fig. 15. Fig. 16.   Fig. 17.   Fig. 18.

cervix. On approximal sides and the curves to the other sides, medium coarse corundum tape and wheels can be used for this purpose.

A root-trimmer or reducer recently devised by Dr. W. S. How can be used in combination with or in place of the appliances before described. It is set in a cone-socket handle. The pin of the center shaft is inserted in the opening in the end of the root and the scraper point rotated around the periphery underneath the gum margin (Fig. 18). The inward spring of the flat scraper shank causes the point to bear hard against the root, while following its outline closely. The root end is reduced without change of contour, while its taper is reversed, so that the greatest diameter is found at the portion acted on by the extreme end of the scraper point (A, Fig. 18). The milled nuts permit the scraper to be adjusted to different sized roots.

In pulpless teeth, the use of excising forceps should be avoided unless the parts admit of it without serious shock to the root. The best plan is to make a succession of holes across the portion to be removed with a spear-shaped drill, and then cut between the holes with a fissure-bur or corundum disk, which will permit of easy removal of the part (Fig. 19).

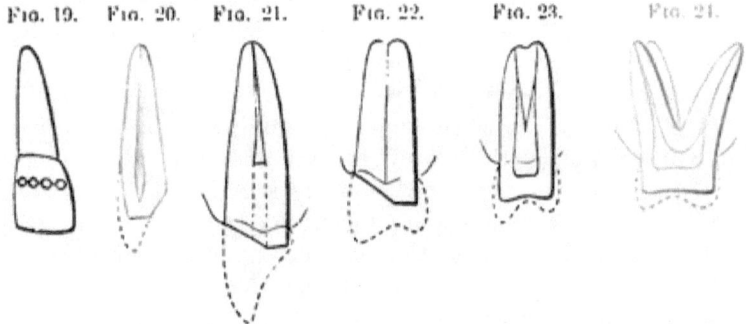

Fig. 19.   Fig. 20.   Fig. 21.   Fig. 22.   Fig. 23.   Fig. 24.

In preparing incisors and cuspids for gold collar crowns with porcelain fronts, where the pulp is to be preserved, the labial surface and incisive edge should be ground down as much as possible without exposing the pulp or subjecting it to irritation; the palatal portion at an angle from the cervical border to the incisive edge, enough to level its prominences of contour and form a slight space between it and the antagonizing teeth (Fig. 20). Pulpless incisors and cuspids should be ground to the margin of the gum at the labial portion, and slightly under the margin on the posterior half (Fig. 21). Bicuspids which are to have porcelain fronts are given the same general form (Fig. 22).

Bicuspids and molars with or without pulps, for all-gold crowns, should have as much of the natural crown left as possible, as it offers a form of attachment for the artificial crown which is more secure and more convenient than is attained by any other method (Figs. 23 and 24).

For the porcelain system, incisor, cuspid, and posterior roots are usually ground level with the margin of the gum. The palatal portion is favored in some cases by leaving it a trifle above the margin. The root-canal is shaped to the form of the post or pin so that it shall fit it tightly. (For details see the articles on the Bonwill and Logan crowns.) The occluding edges or surfaces of antagonizing teeth should be removed sufficiently to allow ample space for the artificial crowns or to favor them in the act of occlusion. Corundum or composition wheels or points should be kept wet and cool in these and other operations in the mouth.

Special Preparation of Badly Decayed Teeth or Roots.—The temporary exposure of the end of a root or of the cervical portion of a crown for the purpose of facilitating or simplifying a crowning operation, especially in the adaptation of a collar or band, is effected by inserting in the pulp-chamber or the root-canal a piece of gutta-percha large enough to admit of a portion being brought over against the investing membranes, to compress them for a day or more. Where a secure attachment for the gutta-percha cannot be obtained, a plug of wood should be inserted temporarily in the root, and the gutta-percha held in position by being packed around it. The root can thus be exposed to the border of the alveolar process if desired. In bicuspids and molars, when decay extends up on the cervix farther than will the edge of the artificial crown or the collar, the gums should be pressed up with gutta-percha, the decay removed, retaining-pits for a filling made, and the cavity filled with amalgam shaped to the contour of the tooth (Fig. 25). In incisors and cuspids, when extensive decay has destroyed a portion of the side of the root, a tight-fitting tube made of a metal to which amalgam will readily adhere, and of such size as will admit the pin of the crown, can be inserted up the root-canal and the upper end cemented in with oxyphosphate and the lower with the amalgam forming the filling on the side of the root. In such a case, the pin sup-

Fig. 25.

porting the crown should be tapered at the end, and inserted in the canal as deeply as possible beyond the end of the tube. Additional strength is thus obtained by a distribution of the leverage along the whole line of the root.

When a gold cap-crown is to be adjusted on a badly broken-down tooth or root, a post of silver or iridio-platinum wire should be formed to fit the root-canals as shown in Figs. 26, 27, and 28, with a piece of silver soldered crosswise. The post should then

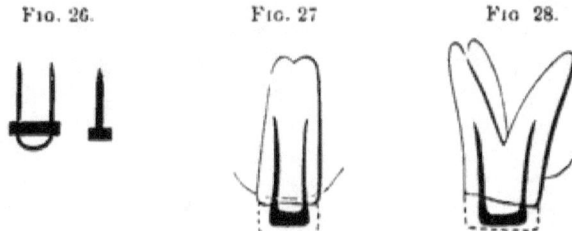

Fig. 26.   Fig. 27.   Fig. 28.

be barbed and the point first fastened in the root with a little oxyphosphate, and the crown portion built down about two-thirds of its length with a quick-setting amalgam. This when hard should be shaped and then roughened to furnish a better attachment for the cement with which the crown is set. Screws can be used as posts to support the amalgam if preferred. As the artificial crown completely covers all the amalgam, the existing prejudice against its use cannot apply to this method.

PART II.

ARTIFICIAL CROWN-WORK.

ARTIFICIAL CROWN-WORK.

ARTIFICIAL crown-work affords extensive facilities for restoring the crowns of natural teeth, and furnishes means of support for bridge-work.

Two general systems are presented,—the porcelain and the gold. The porcelain system includes porcelain crowns with metallic attachments, with or without collars, and the gold system gold crowns with porcelain fronts.

# THE PORCELAIN SYSTEM.

CHAPTER I.

PORCELAIN CROWNS.

Porcelain crown-work is practiced by many dentists almost exclusively, excepting only the occasional insertion of a gold cap-crown on a posterior tooth. The reasons for this are, the intricate character of the construction of gold crowns, and the objection to crowns with bands or collars.

The advocates of the all-porcelain system claim for it natural appearance, restoration of contour, strength, and cleanliness, together with simplicity of construction and easy adaptation and attachment to the root, to which the crown is hermetically sealed.

Porcelain crowns are made in two general styles. In one the crown is attached to the root by a pivot, post, or screw, one end of which is cemented in the root and the other in the crown. Such are the Bonwill, Gates, Foster, How, and Howland crowns. In the other style, one end of the pivot, or post, is baked in the porcelain when the crown is made, and the other end cemented into the root when the crown is adjusted. The Logan, Brown, and new Richmond are of this class.

Special advantages are claimed for each of these several forms of crowns. A general knowledge of the different methods is therefore essential to determine the adaptation of each to the requirements of a case.

The preparatory treatment of the roots respecting the process of grinding, trimming, and shaping is nearly the same for all porcelain crowns. Exposing the end of the root, by pressing the gum away from it with gutta-percha preliminary to the operation, will greatly assist the operator, by enabling him to avoid accidental laceration of the adjoining membranes, and the annoyance attending their bleeding, besides permitting him to carefully study the adjustment and cementation of the crown. Crowns in which the post is cemented will first be described.

THE BONWILL CROWN.

This crown, being one of the first introduced in improved porcelain crown-work, has been very extensively used. The process of its adjustment and insertion is explained in a lengthy article by its inventor and advocate, Dr. W. G. A. Bonwill, from which the following is presented:

"These all-porcelain crowns have three distinctive features: a concave or countersunk base; a triangular opening from the base to a point at or near the cutting-edge of the incisors, the base presenting to the labial surface (at its upper portion this groove is enlarged); a peripheral margin or border resting perfectly flat on the root, the concavity of the base on the palatal side being at a much more acute angle than on the approximal sides. An anchorage is made in the incisors by a depression or undercut between the labial and palatal surfaces, opening on the latter. In the bicuspids and molars the retaining-pits are nearer the grinding-surface.

Fig. 29.   Fig. 30.   Fig. 31.   Fig. 32.   Fig 33.   Fig. 34.   Fig. 35.

Fig. 29.—Sectional view of an incisor crown, from mesial side, showing the undercut at the point opening on palatal surface, the conical base, and the opening from the same to the retaining-grooves, with the exact relations.

Fig. 30.—Palatal view of same tooth. *a* is the external opening for egress of alloy and for packing around the pin. The dotted lines show the recess or undercuts on the mesial and distal sides and near the point for retaining the crown, and its relation with the conical base.

Fig. 31.—Grinding-surface view of a superior molar with the countersunk pin-holes on the buccal and palatal sides.

Fig. 32.—Same view of an inferior molar with the pin-holes on the mesial and distal sides.

Figs. 33 and 34.—Sectional views of a molar and a bicuspid crown, showing the countersinks and their relations with the conical base.

Fig. 35.—Sectional view of an incisor root, showing the retaining-cuts made by the wheel-bur shown in Fig. 42.

"The concave base of the crown prevents the amalgam from escaping under the heavy pressure exerted to force it into position, and in impacting the amalgam and expressing the mercury. It allows of a dense body of material around the metallic pin, giving the equivalent of a pin the whole diameter of the base of the crown. It leaves no joint, the crown and root being continuous. The amalgam is so thoroughly hardened at once by impaction in the double concave of crown and root as to make a very firm operation. It prevents any possibility of the crown's twisting upon the pin and root. In the event of fracture of the crown, the convex surface of amalgam on the root makes

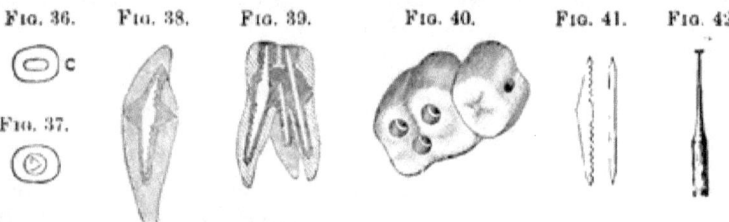

Fig. 36.—End view of a canal prepared for the improved combination-metal pin.

Fig. 37.—End view of same canal as in Fig. 36, prepared for a triangular pin, showing how much more of the mesial and distal surfaces have been cut away from it than in Fig. 36 for the improved pin.

Fig. 38.—Sectional view of an incisor crown and root, with the improved pin in its relative position to each,[1] with the depressions made by wheel-bur.

Fig. 39.—Sectional view of a superior molar, with the large angular pin in palatal root and two square pins in the buccal roots, one being shorter and not passing through the crown.

Fig. 40.—Block of a molar and bicuspid, showing the countersunk holes for pins in the molar, and the hole in the mesial side of the second bicuspid where a pin is alloyed in and set into a decayed cavity in the distal surface of the first bicuspid, being held upon the molar roots and attached to the bicuspid by the alloy.

Fig. 41.—Side and end view of the largest size angular combination-metal pin with the stamped serrations.[2] The square pins are without serrations and double-pointed, made of the same metal and of equal thickness throughout.

Fig. 42.—The smallest-sized wheel-bur for grooving the canals for anchoring the pin and alloy.

[1] The sectional views of the incisor and molar, giving the relative position of the pins in the crowns and roots, should show pins of larger size. The pins as furnished should be filed down but little. It is not absolutely necessary that so many serrations should be made in the canals by the wheel-bur for retaining the amalgam and pin as are shown in the sectional view of the root of an incisor. While no serrations are shown in the roots of the molars, it is understood that all the canals must have the serrations. The square pins in the canals need no serrations. At the point where they occupy the countersink in the crowns, make two or three very slight cuts on the edges with a sharp file. The ends can be left blunt.

[2] These pins are now made without serrations. When amalgam is used for securing them, they become amalgamated and firmly united.

the substitution of a new crown an easy operation. It enables the operator to fit the crown in much less time; it allows a proper position to be given to the pin, with less danger of fracture therefrom; it permits of a larger quantity of amalgam in the crown, and is capable of bearing greater strain; it makes the permanent success of the operation probable, from the fact that it is absolutely jointless, and secures immediate solidity, even while the amalgam is semi-plastic. These crowns are capable of resisting the force of biting or mastication, because they are supported nearly to the cutting-edge or grinding-surface, the triangular opening from the concave base nearly to the cutting-edge allowing the pin to be imbedded in the labial face of the crown where there is the greatest amount of porcelain.

"The amalgam to be used as the medium of union must set quickly and be very hard. Thus far I have found nothing better than the alloys I have specially prepared for this line of work, and, though they are costly, the superior results obtained by their use amply repay the cost. I use No. 1 generally. If mixed thick, it will set so quickly that the operator must work rapidly to prevent its being wasted. In incisor cases I use No. 3 at the gum line and make a close joint.

"In preparing the canal, use first a small-sized, spear-shaped drill, carefully following the natural channel. Then follow with a larger one, taking care not to cut through the root near the apex. On the mesial and distal sides cut away but little, as there is where fractures are most liable to occur. The canal can be very tapering and yet hold the pin. There need be but very little space around the pin. By all means save all the walls of the root possible. The smallest-sized wheel-bur may be used to make an interrupted female thread at various points along the canal to hold the amalgam.

"If the patient exposes the gums much in speaking or smiling, the root may be cut down with the bur or corundum-wheel beyond the free edge to conceal the joint. With bicuspids and molars it is not necessary to go below the gum; a joint well made will not be observed, and the strength of the root will be preserved. If the root is decayed below the gum, after removing the softened parts, fill it with alloy.

"It is not necessary that the face of the root should be flat; it may be either concave or convex, according to indications.

"It is advantageous to take an impression and 'bite' of the root, and make a model and articulation in plaster.

"The crown to be inserted should be inspected closely, as the retaining undercut in the incisors and the depressions in the bicuspids and molars may not be well defined. If not, the crowns are liable to work loose. If the base has been ground off in fitting, the edges should be beveled again to a fine margin with a corundum-point. The crown should be fitted to the root in the mouth, not to the plaster cast. The articulation should be clear, to avoid displacement. The pin should be as large as the previously prepared canal will admit. The pin must in every case be fitted, and in fitting it file only on the plain sides. Leave the end sharp, to offer the least resistance in passing through the amalgam. The end of the pin to be passed into the crown needs very little alteration. The crown being open on the palatal surface of the incisors, permits a blunt-pointed pin to go up to its place. The middle of the pin should not be interfered with if it can be avoided. It is well to cut the pin a little short for incisors, as it may not get pushed entirely up in the root through the amalgam. Small square pins are used in the bifurcated roots of bicuspids and in the buccal roots of molars. They can be sharpened at both ends, but the outer end will not require so much sharpening. The palatal roots of molars will generally take one of the largest thick pins, with one square pin in the largest and most accessible buccal root. Each canal should have a pin, if the canal can be reached and properly prepared to receive it, even though the pin has to be so short as not to pass through the hole in the crown. If it enters the countersunk base it will support the root. The lower molars will require two of the largest-sized pins. As the support of the root is dependent upon the size of the pin and the depth to which it is inserted, single-rooted teeth should have the very largest thick pin. If the root is thin on the mesial and distal sides, the thin, angular pin is to be preferred. Ordinarily these large pins do not have to be bent. If necessary, it had better be done with a hammer, and before the mercury touches them. The pin should have free movement in both root and crown. Should it be discovered that the pin is too long after it has been packed in the root, it can be cut off with sharp forceps, pressing them up against the pin to prevent displacement. The pin can be sharpened subsequently with the corundum-wheel.

"To insure an amalgamation of the pin with the filling, brighten the surface of the former before inserting.

"The roots, crown, and pins being in readiness and arranged on the table, so that no mistake may occur from getting the pin in the wrong position, and the appliances necessary for the operation being at hand, the alloy preferred should be mixed a little thinner than if intended for a filling, especially where the root has a long canal. The shorter the canal, the thicker the amalgam may be mixed. Mix only enough at one time for one root. Put enough amalgam in the canal to nearly fill it, but do not pack it; force a steel pin made for the purpose, of about the same size as the pin, to make way for the easier insertion of the latter. Then grasp the pin with suitable forceps, and carefully but steadily press it up to its destination. If you cannot succeed in doing so, remove it, and again use the steel pin. When in place, use an instrument with a point small enough to pass between the pin and the root, and pack by tamping the amalgam around it. A piece of bibulous paper placed over the point of the instrument will assist materially in carrying the amalgam before it. Before the amalgam has become too hard, replace the crown to determine if the pin is in proper position; if not, it can be crowded to one side or the other with the tamping-tool. Should the pin be found to be rather long, it can be ground off with the corundum-wheel, holding it meanwhile with the forceps. No attempt should be made to bend the pin after it has been amalgamated, for fear of breaking it. If any amalgam has been left, and it is still plastic, it may be packed around the pin at the base of the root, using the bibulous paper as before directed. If not, mix again to complete the operation. Bank up the amalgam on the root high enough to fill the base of the crown. The crown should now be tried on, and forced home with an adjuster adapted to the case, removing the surplus amalgam if too much, or adding if not enough. Remove and dry the crown, and fill up simply the undercut cavity near the cutting-edge if an incisor, or the depressions in the crowns of bicuspids or molars, allowing a very little to extend into the cervical base. Now force it home with the adjuster. It requires considerable force to set one of these crowns according to directions,—a force which cannot be applied with a mallet without danger of loosening or displacing the crown. Steady pressure with slight rotation will carry the crown into place, if the amalgam is not too hard or there is not too much of it. I would advise you not to attempt to set a crown without an adjuster or its equivalent. Free mercury will be squeezed out on the palatal surface, which should be wiped off. Now hold the crown in place with the fingers, with the bibulous paper under the tamping-instrument, and consolidate the amalgam around the point of the pin in the crown, absorbing any free mercury which appears there. The excess of alloy at the joint must now be removed, care being taken to press the crown up while this is being done. The amalgam packed around the pin in the crown on the palatal side should be as stiff as may be to work readily. It is well to leave over some of the first mixing for holding the pin, and this will be about right for consolidating about this point.

"If in a bicuspid or molar crown the pin should come so far through as to interfere with articulation, it may be ground off with the corundum-wheel while the crown is firmly held.

"The case can now be dismissed, with directions for the patient to return the next day, in order to make sure that the articulation is correct and to dress off the joint between the crown and root, which may be done with a small round-headed bur.

"There are some cases in which the root cannot be filled with anything; if in a molar, the pulp-chamber can be relied upon to hold a headed pin or pins. When a tap-hole is required in the root it can be made low down and at an acute angle, and the amalgam packed around the root-canal above the tap.

"Should an artificial crown be broken, another can easily be substituted, by burring off any excess of amalgam, and using fresh amalgam, mixed thin, to allow of ready adjustment.

"Two crowns can be inserted on the root of one large molar with the assistance of the decayed approximal surface of an adjacent tooth (see Fig. 40)."

THE HOW CROWNS AND METHODS.

Fig. 45.   Fig. 46.

These crowns are the invention of Dr. W. Storer How. There are two styles,—four-pin crowns for incisors, cuspids, and bicuspids, and porcelain dovetail crowns for bicuspids and molars. Each form embraces some novel features. Dr. How's methods, being general in application, are used in inserting other forms of crowns.

The following are Dr. How's descriptions and illustrations of his methods and crowns:

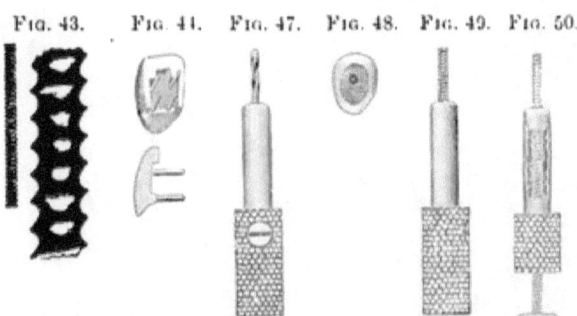

Fig. 43.   Fig. 44.   Fig. 47.   Fig. 48.   Fig. 49.   Fig. 50.

"*The Four-Pin Crown.*—The difficulties and uncertainties in mounting artificial tooth-crowns on roots, by either old or new methods, led me to a careful study of the problem, and resulted in a nearly simultaneous devising of several new forms of crowns and appliances for setting them, as well as a perfected method of performing the operation of fixing a peculiar screw-post (Fig. 43) in a root, and also a novel process of attaching the crown to the post. At present I will describe simply the four-pin crown (Fig. 44) and the successive steps to be taken in mounting it.

"1. When the root is in proper condition for mounting, measure the depth of the canal by means of the canal plugger and its flexible gauge (Fig. 45), and fill the canal at and a short distance from the apex of the root, keeping the gauge at position to show the full length of the canal and also the distance to which it has been filled.

"2. Cut off the root-crown with excising forceps and a round file, down to the gum margin, and with barrel bur No. 241 cut the labial part of the root fairly under the gum without wounding it.

"3. Set gauge on a Gates drill (Fig. 46) to one-half the gauged depth of the canal, and drill to that depth.

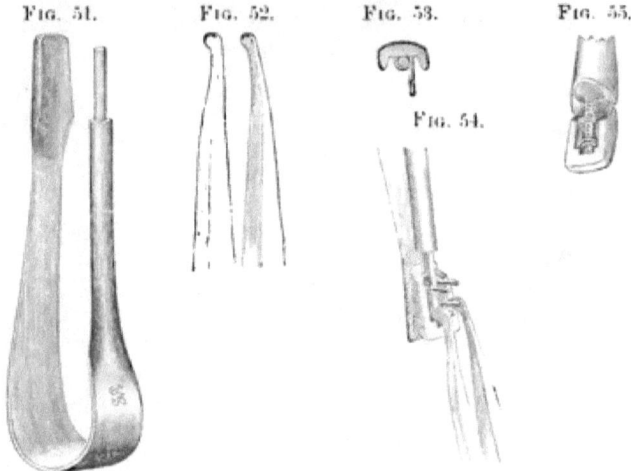

Fig. 51.   Fig. 52.   Fig. 53.   Fig. 55.

Fig. 54.

"4. Set the twist-drill in its chuck (Fig. 47) to project the same length as the Gates drill, and drill the root to exactly that depth.

"5. Enlarge the mouth of the canal one-sixteenth of an inch deep all around to near the margin of the root, as shown in Fig. 48, using square-end fissure-bur No. 59, and then with oval, No. 94, undercut a groove lingually and at the sides.

"6. If the rubber-dam is to be used for a gold or plastic backing, put it now over the root with Hunter's root-clamp, also over the adjacent teeth, and thoroughly dry the canal.

"7. Set the tap in its chuck (Fig. 49) a trifle less in length than the drill, oil it, and carefully tap the root to the gauged depth.

"8. Insert the post in its chuck (Fig. 50) to the exact gauge of the tap, and turn the thumb-screw down hard on the end of the post, then screw the post into the root, release the thumb-screw, unscrew the chuck a half-turn, bend the post until the chuck stands in center line with the adjoining teeth, and unscrew the chuck from the post.

"9. Slit the rubber back from adjacent teeth, tucking the flaps out of the way, so that the occlusion may be tried, the post excised and ground off until the teeth close clear of the post.

"10. Try the crown on the post, and with disk F grind the rib between the neck pins until the crown is labially flush with the root margin, using the disk dry and cutting a little at a time until exactly flush.

"11. Take the crown and place the mandrel (Fig. 51) between the pins just as the post is to be, and with the pliers (Fig. 52) bend the pins carefully over the mandrel, cutting off the pins if too long to be pinched in on the mandrel at the sides, observing that the pin nearest the cutting-edge is first to be bent (Fig. 53), and the opposite pin bent below it on the mandrel, and so with the others (Fig. 54).

"12. Slip the crown over the post, try occlusion, and with the post-chuck bend the post until the crown is properly aligned with the teeth, then with a stump corundum-wheel No. 3 grind the neck of the crown to a close labial fit with the root, fitting only the portion to be concealed by the gum, leaving narrow gaps at the sides to be filled by the backing between crown and root (Fig. 55).

"13. Grind cutting-edge for occlusion and relation to the other teeth, and be sure that the opposing tooth does not strike the crown, the post, or the pins.

"14. Fix the crown on the post by pinching the pins into the screw-threads in the post with the special pliers for that purpose.

"15. Finally, pack the backing of gold, or cement, or amalgam, or Wood's metal,[1] or—for temporary backing while treating abscess—gutta-percha, into all the crevices around the post and behind and under the pins, and between the crown and the root; contour and finish thoroughly, so that no ledge or other imperfection can be found.

"Fig. 56 shows in vertical mid-section an incisor crown mounted on a root; the blackened portions of the backing

---

[1] **Wood's metal** suggested by Prof. J. Taft.

serving to define clearly the locking-hold of the backing on the screw-post, the crown-pins, and the root recess.

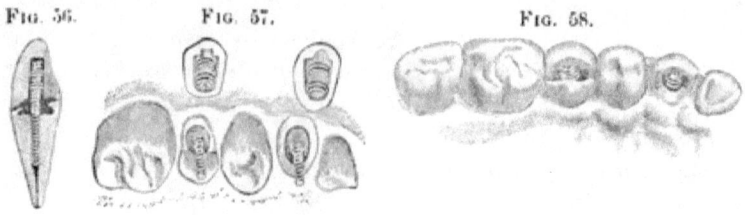

Fig. 56.   Fig. 57.   Fig. 58.

"Fig. 57 shows in perspective a cuspid crown ready to be slipped over its post, and also a cuspid crown ready for its post in the bicuspid root, which has its lingual cusp remaining; and Fig. 58 shows these crowns on their posts awaiting the completing or contour-backing.

"When it is desired to contour the backing of a cuspid crown to form an inner cusp, or to adapt a cuspid or incisor crown for masticating uses, the pins may be twisted together over the mandrel, and again twisted tightly over the post as in Fig. 59; but

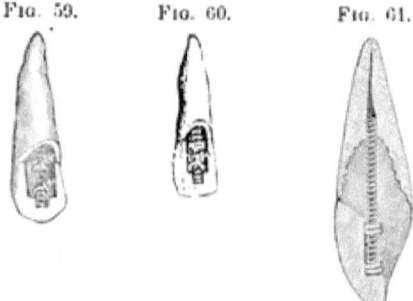

Fig. 59.   Fig. 60.   Fig. 61.

in some cases it may be better to bend the neck-pins, as in Fig. 60, instead of twisting them. In all cases the bent pins are to be pinched quite hard over the mandrel and post, so that the serrations of the pliers will roughen the pins to prevent their being pulled through the backing, which should also be carefully condensed around the pins and post.

"When the root is much decayed, the bottom of the cone-shaped cavity may be drilled and tapped to the depth of a sixteenth of an inch, and the post, thus anchored, may be further secured by cement in the grooved walls of the cavity and around the post (Fig. 61).

"The screw-posts are made of crown metal, an alloy devised for the purpose in order to obtain a stiff post that will permit the cutting of the peculiar and extremely accurate thread formed upon it, and which will not amalgamate or be otherwise affected by any backing-material that may be used. Of course platinum or platinum alloyed with iridium may be employed for posts, but the crown metal is in every way superior.

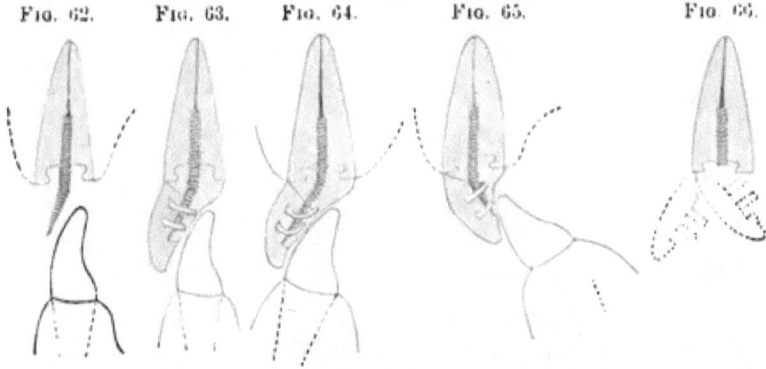

Fig. 62.   Fig. 63.   Fig. 64.   Fig. 65.   Fig. 66.

"There are some cases of a class which has hitherto presented difficulties that may now be easily overcome by grinding the post flat on the crown side after it has been set and bent in the root (Fig. 62), so as to be clear of the occluding tooth; and then the crown-pins may be bent over the reduced post, the crown fitted and ground to clear the opposing tooth (Fig. 63), and the backing added in completion.

"A similar case, in which the opposing tooth and a proper alignment require an oblique bending of the pins, is seen in Fig. 64, while the reverse arrangement of parts is shown in Fig. 65. The crown is thus seen to be adapted to a wide range of adjustments, because its point of contact with the root is at the labial portion of the neck, on which as on a hinge the crown may be swung out or in (Fig. 66, dotted lines), over an arc of at least sixty degrees, at any point of which it may be quickly and firmly fixed. The labio-cervical junction is made just under the gingival margin, and I usually interpose a thin layer of cement, amalgam, or gutta-percha, or a narrow ribbon or several large blocks of soft gold; the joint always to be made carefully smooth, and hid from view under the free margins of the gums."

The Porcelain Dovetail Tooth-Crown.—These crowns are designed for the roots of bicuspids and molars only, and the process of mounting them may be very briefly described.

"Fig. 67 shows the roots of an inferior molar after the apical portions have been filled, the neck recessed, the canals drilled and tapped, and two How screw-posts firmly fixed therein, the ends of the posts having been pinched towards each other by means of a pair of pliers, so that they will go through the central opening in the crown (Fig. 68). This opening is of a dove-

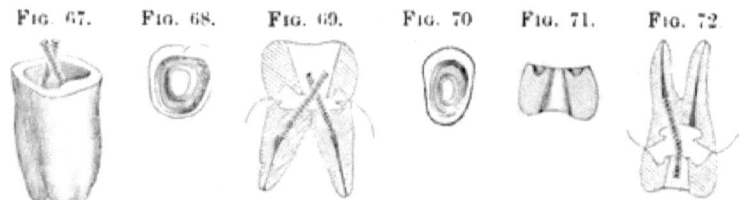

FIG. 67.   FIG. 68.   FIG. 69.   FIG. 70.   FIG. 71.   FIG. 72.

tail form, as shown in cross-section by Fig. 69, where the crown is seen in place over the posts on the root. It is thus made obvious that the crown may be easily put on and off the root in the process of fitting the crown-neck to the root-neck, and also that, for occlusion, the crown may be ground low on any or all sides without destroying the dovetail function of the central cavity. When the fitting is completed, and the crown cut so short as to be $\frac{1}{32}$ of an inch distant from the occluding tooth, amalgam is packed into the neck recess, around the posts, and thinly over the cervical margin of the root, the crown put in place, and, with thumb pressure, firmly seated. Then test the occlusion, and complete the operation by packing amalgam into the crown opening, which will permit the forcing of the amalgam in all directions, to insure a firm base for the crown, and its secure dovetail attachment to the posts, as shown by Fig. 69.

"The bicuspid crown (Figs. 70 and 71) is similarly mounted, as may be seen in Fig. 72, cross-section; the same crown and root being shown in contour by Fig. 73. In some instances this bicuspid crown may, like the Foster crown, be secured by a headed screw, as shown in Fig. 74. The root having been

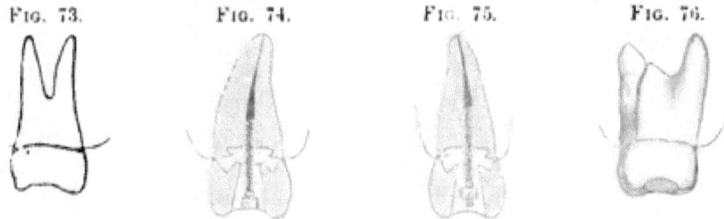

FIG. 73.   FIG. 74.   FIG. 75.   FIG. 76.

drilled and tapped and recessed, and the crown properly fitted and articulated, the screw is put through the crown, amalgam packed in the crown-groove and around the screw, which is then inserted in the root, and the crown pressed hard into its place. The screw is then turned into the position shown in Fig. 74, thus compressing the amalgam or cement in both recess and groove, after which the screw-head may be covered with amalgam, cement, or gold, as desired.

FIG. 77.

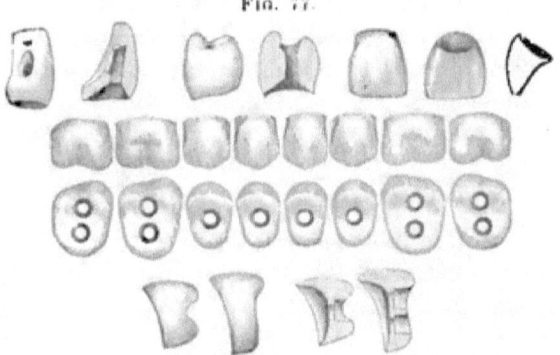

" As a preferable mode, however, the screw-post may first be fixed in the root, the crown adjusted over the post, amalgam packed on the root and around the post, the crown seated firmly, more amalgam packed in the crown cavity around the post, and then a nut screwed on the post, as shown in Fig. 75. In all the sectional cuts cement, amalgam, or gutta-percha is to be understood as filling the cavities in the conjoined roots and crowns.

"Fig. 76 shows in contour a dovetailed crown mounted on a superior molar root in the manner shown by Fig. 69. It is obvious that the crown of Fig. 69 might be ground quite down to the post-ends, and yet be firmly held by the dovetail sides of the central cavity."

THE GATES CROWN.

The Gates crown has become identified with the Bonwill, owing to its similarity (Fig. 77). It is usually attached to the root by a metallic screw manufactured for the purpose, such as is illustrated in Fig. 78, instead of the Bonwill pin.[1] The screw is first inserted in the root and the amalgam packed around it. In nearly all roots, at a reasonable distance up the canal, a suitable place for fastening the end of the screw can be found. Too much force must not be applied in its insertion, as a root is easily split. In bicuspid and molar crowns nuts are used on the screws, which fit slots in the grinding-surface of the porcelain. They are screwed into the amalgam or cement, and covered with it in the process of cementation of the crown.

When it is desirable or necessary to construct a screw for a special case, it should be made of iridio-platinum wire (as this alloy, being hard, will well maintain the edge of the thread). Gold is unsuitable, owing to the action upon it of the mercury in the amalgam, even though the amalgam is used "dry."

In forming a screw a coarse-thread screw-plate should be used. The threads of most screws are cut too fine.

THE FOSTER CROWN.

The Foster crown (Fig. 79), which in general form is similar

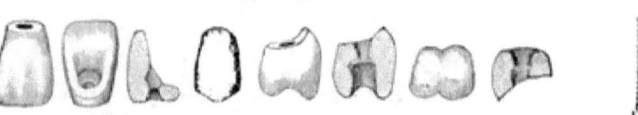

Fig. 79.   Fig. 80.

to a crown introduced by Dr. H. Lawrence, of Philadelphia, in 1849, also resembles the Bonwill, but has less concavity at the

---

[1] Dr. Gates originally used a double-ended oval-shaped screw. The How screws (see page 55) manufactured by The S. S. White Dental Manufacturing Company are applicable to these crowns.

base. The crown is attached to the root by a headed screw (Fig. 80) or a screw with a nut, instead of the Bonwill pin.

The How screws and instruments (Fig. 81) are best adapted for use with these crowns.

THE HOWLAND CROWN.

This crown, which is similar in principle to one originally introduced by Dr. C. H. Mack, was devised by Dr. S. F. Howland. It is attached like the How dovetail crown, with screws that are first inserted in the root. It is used mostly on bicuspid and molar roots, and consists of a hollow porcelain crown, with a cavity in the crown sufficiently large to admit the screw or pins, and, when necessary, a small portion of the root (Fig. 82).

The method of setting this crown, as described by Dr. Howland, is to shorten the root even with the gum with a stump file; fit the crown to the root; enlarge the root-canal so that a threaded pin of proper size will pass in easily, partially fill the canal with zinc phosphate, and press the pin to its place with pliers. The crown should then be filled with zinc phosphate and pressed to its place, care being taken to hold it in position until the cement sets (Fig. 83). If any operator distrusts the ability of zinc phosphate to make a perfect joint, a small quantity of silver amalgam or gutta-percha can be used to advantage.

FIG. 81.

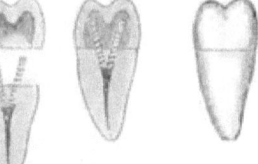

FIG. 82.    FIG. 83.   FIG. 84.

This crown is strong, and during its test of several years none have broken so far as known. The mode of fastening is strong, and has the advantage of two pins when set on a bicuspid having two root-canals. It is simple, and when set—no metal being in sight—it is a perfect imitation of the natural tooth (Fig. 84).

THE LOGAN, BROWN, AND NEW RICHMOND CROWNS.

These crowns have their platinum posts or pins baked in the body of the porcelain. In the Logan crown the base is made concave, to facilitate its adjustment to the end of the root, and to give the cement a more reliable form. The base of the Brown crown is convex, and that of the new Richmond V-shaped, from mesial to distal side.

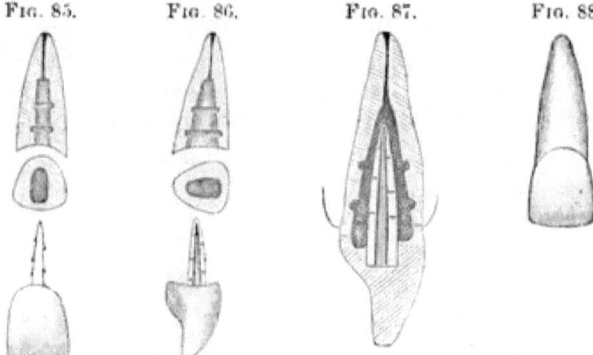

Fig. 85.   Fig. 86.   Fig. 87.   Fig. 88.

The preparation of a root for each of these crowns is, in general, similar to that for the Bonwill crown. The root-canal is enlarged, and shaped so that the post, if possible, at least at its point, will fit tightly. Gutta-percha or oxyphosphate is used for cementing in preference to amalgam.

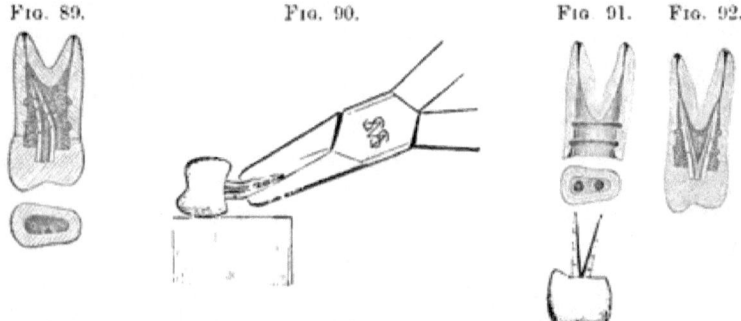

Fig. 89.   Fig. 90.   Fig 91.   Fig. 92.

The Logan Crown.—The Logan crown, now so extensively used, is the invention of Dr. M. L. Logan. The method of mounting is explained in all its details in the following article by Dr. W. S. How:

"Fig. 85 shows a superior right central root, an end appearance of the same, and a Logan crown, front view. Fig. 86 exhibits, at a right angle to the plane of the first figure, the same root, its end, and the Logan crown, side view. In both figures the root-canal is supposed to have been first drilled to a gauged depth with an engine twist-drill, No. 154, and then enlarged by means of a fissure-bur, No. 70, to the tapering form shown:

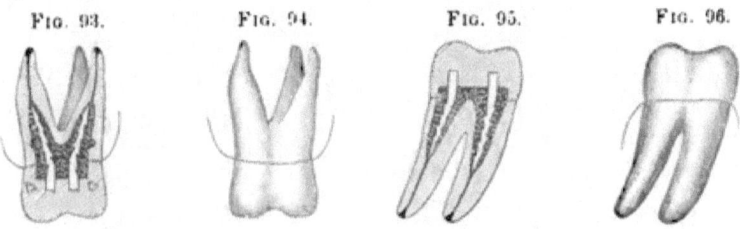

Fig. 93.   Fig. 94.   Fig. 95.   Fig. 96.

the walls being subsequently grooved with an oval bur, No. 90. The enlarged section, Fig. 87, shows the crown adjusted on the root by means of cement or gutta-percha, which surrounds the post and fills all the spaces in the root and crown. Fig. 88 shows the completed crown. Fig. 89 exhibits a bifurcated bicus-

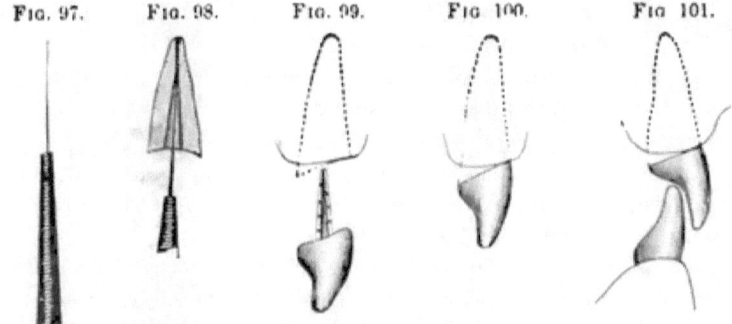

Fig. 97.   Fig. 98.   Fig. 99.   Fig. 100.   Fig. 101.

pid root, its end appearance, and a Logan crown adjusted to the root. Fig. 90 illustrates the best manner of bending the post. Fig. 91 shows a split post, and its adaptation to a bifurcated bicuspid root is seen in Fig. 92. Figs. 93 and 94 exhibit the mode of mounting the Logan crown on a superior molar root, and Figs. 95 and 96 the same crown in its relations to an inferior molar root.

"The preceding figures clearly present to the mind's eye of the expert dentist the essential features of the Logan crown and the method of mounting it.

"The details are as follows: In every instance where a root is deemed ready to receive its filling, it should first be measured through its canal from the cervical opening to the apical foramen, and this may be accurately done with a gauge adjustable on a delicate canal-explorer (Fig. 97). The same device serves to measure the distance from the apex to which the canal should then be filled (Fig. 98). It also gauges the depth to which the drill may be carried. The proper degree of enlargement from the bottom of the drilled hole will, of course, depend on the observed size and character of the root. Every dentist should familiarize himself with generic tooth-forms, so that when the length of an incisor, cuspid, or other tooth-root is known, he can so nearly determine its hidden outlines as to form with precision a corresponding enlargement of the root-canal, such as is shown by the several cuts. For preparing the roots, the Ottolengui root-reamers (Fig. 102) and facers (Fig. 103) are very desirable instruments. The reamers are made in three sizes to correspond with the Logan pins. With a root-reamer of the appropriate size, the root-canal is enlarged to fit the pin along its whole length, and so hold the crown firmly *independently* of the cement. With a root-facer a labial slope is given to the root-end, so that the crown neck shall fit under the edge of the gum. Fig. 104 shows the method and its result, and the cross-section shows how the cement incases the pin. The suitable preparation of the bifurcated roots of some bicuspids and of all the molars is a matter involving difficulties of an unusual character and requiring good judgment. The feasibility of splitting the post of a Logan crown to adapt it to the bifurcated root of a bicuspid is shown by Figs. 91 and 92. This example directs attention to the peculiar shape of the post, in which there is effected such a distribution of the metal that its greatest strength is in the

Fig. 102.

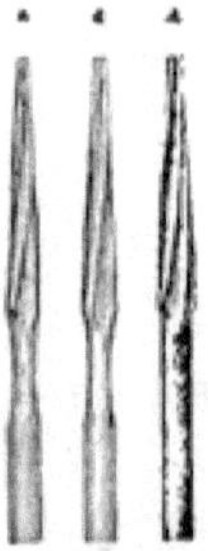

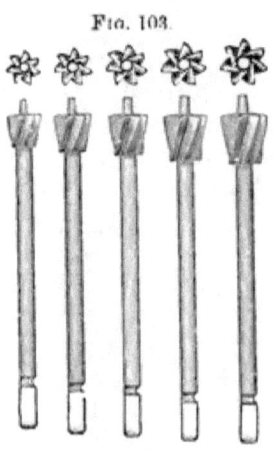

Fig. 103.

line of the greatest stress that will in use be brought to bear on the crown, while the least metal is found at the point of the least strain; the applied part of the post being in outline nearly correspondent to that of the root itself. The root-canal is likewise conformably enlarged to receive the largest and stiffest post which the size and shape of the root will permit.

"The fitting of a Logan crown to a root may be done with a wet stump-wheel in the engine hand-piece. A safe-side crown corundum-wheel (Fig. 105) can be used in the same manner. It also affords the greatest facility for the slight touches required to abrade the thin cervical borders of the crown, which may by this means be done without encroachment on the post.

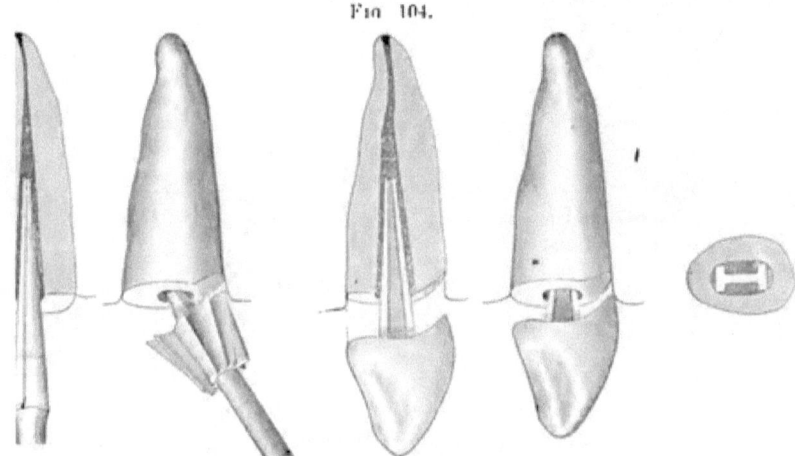

Fig. 104.

"By the old method of adapting pivot-teeth to roots, the close fitting of the crown precluded the use of a plastic packing, because its thinness over the surface of the joint made the packing liable to break loose under the shock and strain of use. The recess in the Logan crown obviates this defect by providing a receptacle for a considerable interior body of cement that will be deep enough to be self-sustaining internally, and yet allow the peripheral portions of the root and crown to approach each other so closely that, though only a film of packing remain, it will still be strong enough to insure the persistent tightness of the joint. This annular boss if formed of amalgam also adds strength in some cases to the mount.

"When enough of the natural crown remains, it is well to leave standing some of the palatal portion, and cut the root under the gum margin at only the labial part, as shown by Fig. 99. The safe-side crown wheel is especially useful in such cases (Fig. 106). Thus the labial joining of the root and crown will be concealed, and the other parts of the joint will be accessible for finishing and keeping clean (Fig. 100). The Logan crown may be ground until a large part shall have been removed for adaptation to the occluding tooth or teeth without seriously impairing its strength (Fig. 101). This crown also in such cases maintains the translucency which is one of its peculiar ex-

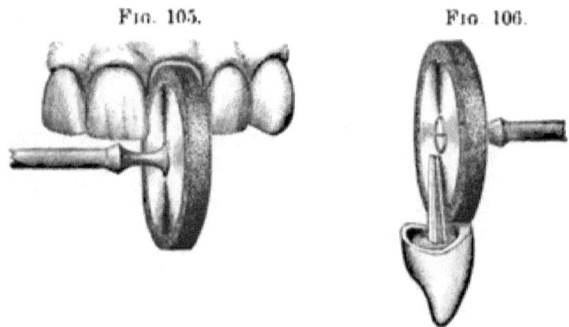

Fig. 105.   Fig. 106.

cellences, owing to its solid porcelain body, and the absence of a metallic backing or an interior largely filled with cement or amalgam.

"The distal buccal root of the natural superior molar is nearly always too small to receive a post of any useful diameter, and therefore the Logan superior molar crown has but two posts, which like those of the inferior molar crown are square, and thus may be easily barbed, as may also the ribbed posts of the crowns for the anterior tooth-roots. These posts are large enough in all the Logan crowns to answer in any given case, and can of course be easily reduced to suit thin or short roots.

"Any of the cements or amalgams may be used in fixing these crowns, but good gutta-percha, softened at a low heat and quickly wrapped around the heated crown-post, which is at once seated in the root, forms the best mounting medium, and has the great advantage of permitting a readjustment, or, if need be, the ready removal of the crown by grasping it with a pair of hot pliers or forceps, and holding it until the gutta-percha is sufficiently softened."

The Brown Crown.—Fig. 107 is a lateral view of a porcelain crown, with an iridio-platinum pin baked in position, invented by Dr. E. Parmly Brown. The pin has great strength at the neck of the tooth, where the strain is heaviest, and this strength is further increased by extending the porcelain up on to the pin.

Fig. 108 is a front view of the same crown, showing by the dotted lines the shape of the pin and the position which it occupies in the crown.

The pin is flattened laterally, affording a strong hold in the porcelain without bringing the pin too near the surface in thin

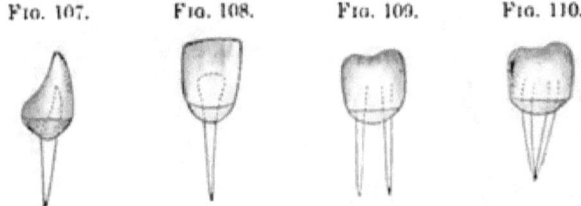

Fig. 107. Fig. 108. Fig. 109. Fig. 110.

teeth, while it also permits alteration of the palatal surface of the crown in a close "bite" without risk of weakening the body.

Fig. 109 is a view of the bicuspid crown, in which a pin is provided for each root of two-rooted bicuspids.

Fig. 110 is a view of a bicuspid crown with the two pins pressed together, forming a single pin of great strength for a tooth with only one root.

The double pin in the bicuspids prevents the gradual loosening of the crown by the rotary movement of the jaws in mastication, which, acting on the two cusps, exerts such leverage as to sometimes turn and break down ordinary crowns where only one pin is used.

The roots are ground concave to fit the crowns with corundum-points or a Willard countersink bur, and close joints are made well under the gum, the pins being set with oxyphosphate cement. The canal should be enlarged enough only to admit the pin, which should fit snugly throughout its entire length, the better to distribute the leverage exerted by the crown, and thus directly to increase the strength of the attachment. (For process of cementation, see article on "Insertion and Cementation.")

The New Richmond Crown.—To illustrate and describe the method of mounting this crown, a superior left central incisor root will serve as a typical case, and its projecting end is to be shaped as seen in Figs. 111 and 112. This can be rapidly done with a narrow safe-sided flat or square file, the angles of the slopes being such that the gum on the labial and palatal aspects will not interfere with nor be disturbed by this preliminary work, as the root end is not, in this operation, to be cut quite down to

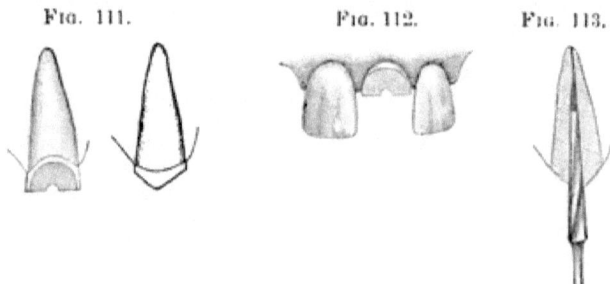

Fig. 111.   Fig. 112.   Fig. 113.

the gum. An Ottolengui root-reamer No. 2 is then employed to bore out the root to receive the crown-post, which is of the same size and shape as the Logan crown-post for a central incisor.

The sectional view (Fig. 113) shows the relation of the reamer to the root. The new Richmond crown (Fig. 114) is then tried on the root (Fig. 115), and its position relative to the adjacent and occluding teeth noted. If the cutting-edge of the crown is to be brought out for alignment with its neighbors, the root can be drilled a little deeper, and the reamer pressed outward as it revolves to cut the labial wall of the cavity. The palatal root-slope must then be filed to make the V correspond to the changed inclination of the crown.

Thus, by alternate trial and reaming and filing, the crown may be fitted to the root and adjusted in its relations until the post has a close, solid bearing against the labial and palatal walls of the enlarged pulp-chamber, and the crown-slopes are separated from the root-slopes by the thickness of a sheet of heavy writing-paper. This space can be accurately gauged, and the root-slopes conformed to the crown-slopes by warming the crown and putting on its slopes a little gutta-percha, so that an impression of the root-end may be taken, and the root-slopes dressed with a file until the film of gutta-percha proves to be of equal thinness on both slopes.

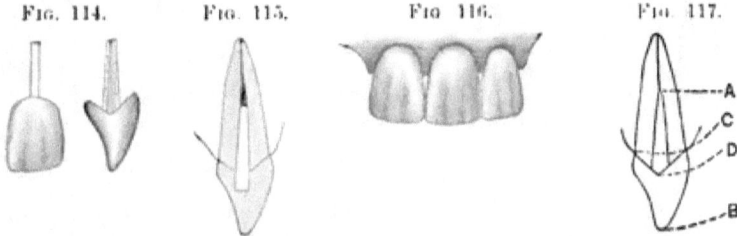

Fig. 114.   Fig. 115.   Fig. 116.   Fig. 117.

To permanently attach the crown, Dr. Richmond usually takes a thin, perforated disk of gutta-percha, pushes the post through it, warms the crown, presses it into place, and when cooled removes it, and with a sharp knife trims away the gutta-percha close to the crown-neck. He then warms the crown, puts a very little oxyphosphate cement on the post, and presses the crown home. Fig. 116 shows the completed crown.

The obvious advantages of the device are the readiness with which the slopes of the root-end may be shaped with a file; the facility with which these slopes may be given any angle to set the crown out or in at the base or at the cutting-edge, or to give it a twist on its axis; the certainty that, once adjusted, the final setting will exactly reproduce the adjustment; the assurance that in use the crown will not be turned on its axis,—a most common cause of the loosening of artificial crowns; the firmness of its resistance to outward thrust in the act of biting. This is made apparent by Fig. 117, wherein it will be seen that in an outward movement the crown B would rock upon A as a pivot. The dotted line D shows how the crown-slope is resisted by the root-slope, which extends so far towards the incisive edge that a much firmer support is given to the crown than if the resistance should be, as it usually is, on the line of the gingival margin C.

For roots that have become wasted below the gum-surface it is not suitable, except in such cases as are decayed under the labial or palatal gum-margin only, but have yet projecting the approximal portions of the crown (Fig. 118).

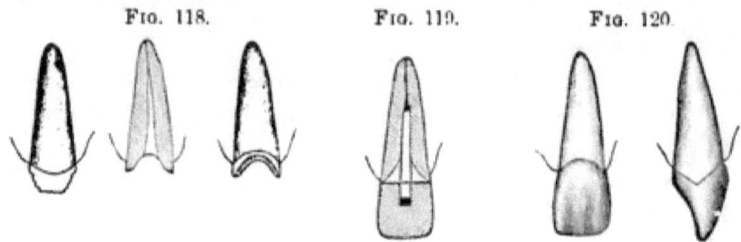

Fig. 118. Fig. 119. Fig. 120.

The sectional view (Fig. 119) and the perspective plan views (Fig. 120) illustrate the manner of mounting these crowns on this class of roots. The finished crown appears as in Fig. 120.

The cases for which this crown seems specially adapted are such as have some considerable portion of the natural crown remaining.

REMARKS ON THE USE OF PORCELAIN CROWNS.

Porcelain crowns have some decided advantages. They are especially useful in many cases where an inexpensive or easily adjusted crown is required; or where some pathological condition limits the probable durability, or permanency, of any operation. In the insertion of porcelain crowns, the removal of the whole or a part of the natural crown, which could be utilized to some extent as a foundation by other systems, has given rise to various objections. If the natural crown is entirely cut away, the pin, or post, upon which almost the entire support of the artificial crown is thrown, acts like a lever in the root-canal, there being no band or brace to relieve the strain. With the whole force of mastication bearing directly upon these pulpless roots, whose disintegration is slowly but constantly progressing, the inevitable result can well be conjectured. Sooner or later they are fractured, and their usefulness as a foundation ended. Then, again, if porcelain crowns are attached with amalgam, the discoloration of the line of union with the root, if subsequently exposed, is disfiguring. These facts, and the lack of strength incident to some forms of construction, are the principal objections urged against porcelain crowns as ordinarily inserted.

CHAPTER II.

PORCELAIN CROWN WITH GOLD COLLAR ATTACHMENT.

A GOLD collar, either seamless or soldered, can be used advantageously in combination with many of the porcelain crowns.

The root having been properly prepared, a collar is adjusted and adapted to it the same as for a gold collar crown (see page 84). The porcelain crown, the base of which should be fully as large as the end of the root, is then ground even with the cervical walls, and fitted into the collar, which should be trimmed and burnished to the form of the crown. Dr. Townsend's fusible metal die, used in the following manner, facilitates the application of a collar to a Logan crown (Fig. 121). Enlarge the root-canal to receive the Logan pin. Grind a Logan crown to fit, and articulate it. Construct a band of No. 30 gold (or of No. 32 crown-metal, which is better) wide enough to project beyond the end of the root say $\frac{3}{32}$ of an inch. Cut a wooden peg about an inch long and taper one end of it to the general size and shape of the pin in the Logan crown. Place the band on the root, insert the peg in the canal, and fill up the band with Melotte's moldine or with stiff putty, pressing it closely about the peg. Remove all together and, holding the die over the flame of an alcohol lamp to melt the fusible metal, place them—the band, peg, and moldine, in the same relative positions they occupied in the root—on the die, with the pin in the socket, and press down until the moldine rests on the surface of the molten fusible metal.

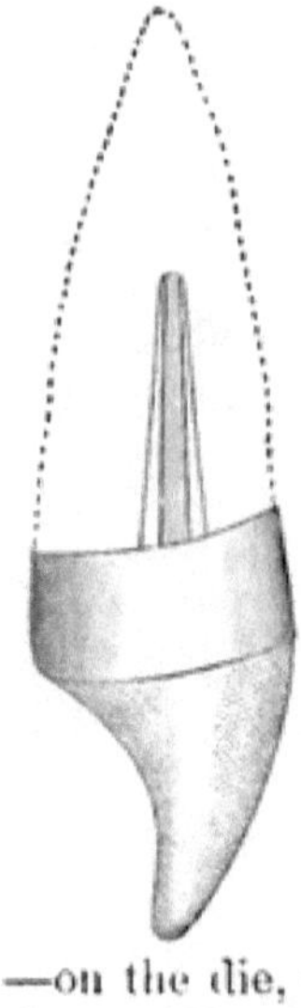

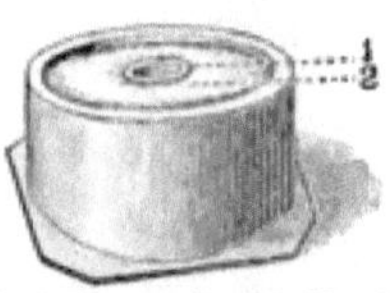

FIG. 121.

1, Socket. 2, Fusible Metal.

Chill; in cooling, the fusible metal takes a firm hold on the lower edge of the gold band, holding it securely in place during the remainder of the operation. Remove the peg and the moldine, and with a wooden mallet drive the Logan crown into the band until the porcelain rests upon the fusible metal. Burnish the band smoothly about the crown. When the gold is perfectly adjusted to the porcelain, melt the fusible metal to release the band and crown.

If the work has been carefully done, the crown with its band will then be ready to be set, as the articulation and fit will not have been disturbed.

Enough of the collar should be trimmed away at the labial portion to prevent too conspicuous exposure of the gold (Fig. 122).

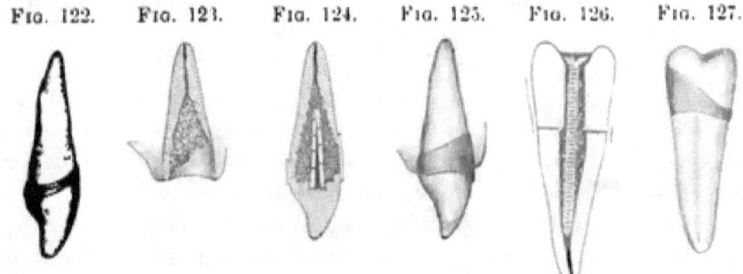

FIG. 122.   FIG. 123.   FIG. 124.   FIG. 125.   FIG. 126.   FIG. 127.

This collar combination is available in very difficult cases, as, for instance, when a root is decayed upon one side far beneath the gum, as seen in Fig. 123.

Such an operation, when completed, would appear in vertical section like Fig. 124, and a view in perspective would resemble Fig. 125. The collar is also very useful wherever the root and crown are not made flush and smooth at every point, as, if practicable, they should always be.

Dr. E. C. Kirk, of Philadelphia, uses for bicuspids and molars a Foster crown in combination with a collar (Fig. 126). A collar is made, and, on being fitted to the root, is cut narrow on the labial side, and left wide on the lingual, so that it shall extend nearly to the cusp of the crown when finished (Fig. 127). (The seamless gold collars are well suited for application to this style of crown.)

The crown selected should have a somewhat greater circumference at the base than the collar, so that when ground down somewhat conically on its lingual and approximal surfaces, it can be tightly adjusted to the collar. If a crown smaller than the collar is used, a tight joint cannot be made. The screw is fitted so that it shall hold the crown in proper relations with the root. The screw and crown are then removed, the parts dried, and the root-canal filled with a slow-setting oxyphosphate cement, mixed thin. The crown is then pressed into its position, the surplus cement flowing through the opening in the porcelain and filling up any interstices around or between the band, the root, and the crown. The screw is then driven into position, and when the cement is set perfectly hard the head of the screw or the nut on it is notched to form a retaining-pit, and the countersink of the crown filled with gold.

Dr. C. S. W. Baldwin, of New York, caps the root and attaches a Logan crown in the following manner:

First, the root is shaped, the outer margin being beveled about the thickness of the gold used, to afford regular sides for close adaptation of the caps. Then an impression is taken and a die made in the gold seamless cap method. To strike up the cap, place No. 32 gauge gold plate on a cushion of lead, holding the die firmly on the gold where you wish to produce the cap, and strike until the required depth is secured before removing it. This drives the gold and die into the lead, forming a female die and a perfect-fitting cap at once, in less time than is occupied in describing the process. Trim the edges to fit the festoon of the gum, and drill a hole from the inner side for the pin, leaving the raggedness made by drilling to catch in the cement. Place the cap on the root and fit the porcelain crown accurately to it in the desired occlusion and position. A Logan crown can, with little grinding, be made to do good service (Fig. 128). A crown having the H-shaped pin, but square on the edge, like some of the early patterns of Logan or Bonwill crowns, would reduce the time of setting and give best results. Having polished the edges of the cap, the crown may be conveniently adjusted as follows: Place oxyphosphate cement in the countersunk portion of the porcelain, and in the canal only enough

PORCELAIN CROWN WITH GOLD COLLAR ATTACHMENT. 77

cement, of creamy consistence, to fill it, as the pressure required to force out the surplus under the edges of the cap destroys many nicely adjusted crowns, leaving bulging irritants instead of smooth supports. If proper attention has been given to fitting crown and root, all will come nicely to place, but in some cases of difficult adjustment it may be necessary to cement the crown to the cap before fastening the pin in the root (Fig. 129).

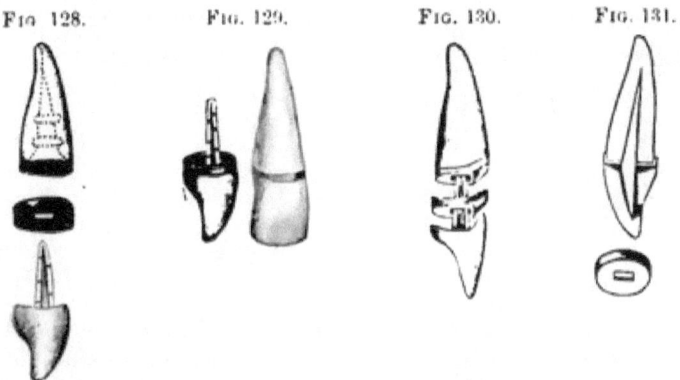

Fig. 128.   Fig. 129.   Fig. 130.   Fig. 131.

In most cases the gold band will be invisible and below the free margin of the gum. Cases may occur where the anterior teeth are prominent, and it will be necessary to cut away the top of the cap in front, allowing the porcelain to come directly in contact with the root, the band going deeper than in ordinary cases, which prevents the appearance of gold (Fig. 130).

Dr. Bonwill's plan is to cap the tooth with a platinum or gold cap having a slot, into which the pin passes as it is slipped on the root (Fig. 131). The crown is then fastened on with amalgam in the usual manner.

CHAPTER III.

THE WESTON CROWN.

Dr. HENRY WESTON's method of crowning is described by him as follows:

"This crown represents on its labial surface the ordinary plate tooth, the lingual or palatal surface being concaved in such a manner as to afford the largest amount of working room without impairing its strength at any point. The pins are so imbedded in the thickest part of the crown that it is not liable to be weakened by grinding. Fig. 132 represents the crown.

FIG. 132.   FIG. 133.    FIG. 134.    FIG. 135.    FIG. 136.    FIG. 137.

"The pin is made of hard platinum and iridium, and is spear-shaped and notched on both edges to give firmness to its anchorage. The backing is of the same metal and strongly soldered to the pin. Fig. 133 represents the pin.

"The preparation of the root for the reception of the crown consists in the removal of that part of the tooth which is to be replaced by the porcelain. Fig. 134 shows the root at this stage.

"The canal must be sufficiently enlarged to allow space for packing gold or amalgam securely about the pin, and the sides reamed (Fig. 135).

"The grinding of the crown to the root requires but little labor, and the fitting may be done directly on the natural root, or on a cast taken after the root has been prepared. This must always be done with the utmost neatness and precision. When fitting, the crown can be held in position by a cone of wax inserted in the root. Next comes the adjustment of the pin to the root of the crown. After punching holes in the backing to receive the pins of the porcelain crown, bend the pin with small flat-nosed pliers, so that when in its proper position there shall be a uniform space around its entire surface.

"Secure the tooth and pin together with a cement of resin and wax, invest in plaster and sand, and solder with fine gold solder.

"It has been my practice of late years when preparing the root to leave just a line of enamel around its entire circumference, thus securing a joint clear of the free edge of the gum, especially when gold is used. If gold is to be used as the attachment, the rubber-dam is indispensable. When amalgam or cement is used, the rubber-dam may be dispensed with by those who prefer other methods of keeping out moisture. When gold is to be used, the root having been previously properly treated, and everything in readiness and the rubber-dam in place, put upon the point of the pin a pellet of phosphate or oxychloride of zinc, the size of a No. 7 or No. 8 excavating bur; now press the pin and crown carefully to their exact position into and upon the root, and with a delicate but blunt-pointed instrument, thin enough to reach the end of the canal, pack the cement firmly about the pin. The object in using the cement is to secure the pin in its place during the first introduction of the gold. By using the hot-air syringe, the cement will harden in two minutes. Close the opening of the canal about the pin with a rope of bibulous paper, and attach the crown to the root and adjoining teeth on either side with soft wax; see that the joints are exact in every particular, as after the next step mistakes are not easily remedied.

"Paint the joint from the labial side with cement mixed to the consistence of cream. Cover the labial surfaces extending over the cutting-edges of the porcelain crown and adjoining teeth to the thickness of three-eighths of an inch with carefully mixed impression plaster. When hardened, the plaster may be cut from the cutting-edge of the crown, and the wax and the paper removed. Now paint the palatal sides of the joint with the cement mixed to a cream-like consistence, applied with flattened root-canal pluggers. Harden with hot air, and the case is ready for the gold.

"When filled and finished, the exposed dentine and enamel are all covered with cohesive gold and porcelain; the result will be a fac-simile of Figs. 136 and 137.

"If preferred, the cement in the labial joint may be dispensed with, and the space between the crown and the root carefully filled with gold after the removal of the plaster, covering all of the exposed root and showing only a fine line of gold at the margin of the gum.

"Where amalgam is used exclusively for the attachment, the greatest accuracy should be observed in the proper articulation of the crown before introducing or packing it and in removing carefully all excess, and the patient should be cautioned against biting on the crown until the following day."

CHAPTER IV.

PORCELAIN CROWNS WITH RUBBER OR VULCANITE ATTACHMENT.

Fig. 138 illustrates the formation of such a crown. The root, when prepared, extends at the palatal side a little below the line of the gum (A) at the point B. A plate tooth (C) is ground and fitted to the root. An iridio-platinum post is then fitted to the root, flattened slightly and bent at D, and riveted to the tooth. The proper alignments of the tooth and post to the root are then obtained, and they are invested and the post soldered and strengthened at the point D. The backing is then grooved and notched slightly, wax applied, the crown adjusted to the root, and the wax shaped so as to form a foundation and overlapping edge at the palatal portion (E). The crown is then removed, invested in a flask, packed with rubber, and vulcanized. In trimming and finishing, the rubber is allowed to form a partial band or collar around the palatal portion of the root, where it will not show. It is then cemented on to the root with oxyphosphate.

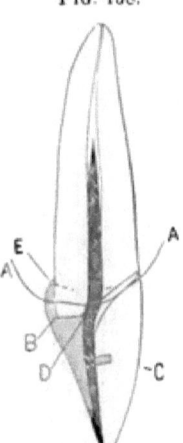

Fig. 138.

The Weston crown and the four-pin How crowns have advantages over ordinary plate teeth in the construction of this style of work.

THE GOLD SYSTEM.

CHAPTER V.

PORCELAIN AND GOLD CROWN WITHOUT A COLLAR.

The root of a cuspid will be taken as a typical case to illustrate the construction of this style of crown.

The end of the root is prepared the same as for a porcelain crown (Fig. 139). The root-canal is then uniformly enlarged a reasonable distance up, with a drill which will tightly fit the opening. Into the canal, gauging its full diameter, is fitted a

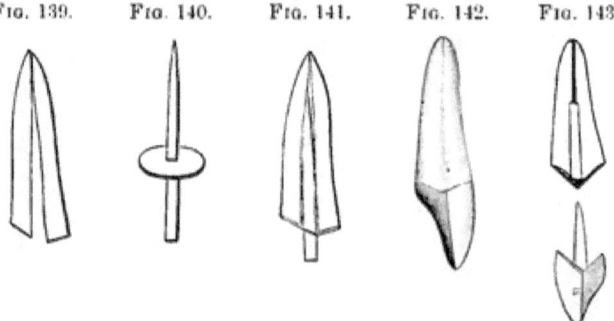

Fig. 139. Fig. 140. Fig. 141. Fig. 142. Fig. 143.

piece of iridio-platinum wire, tapered off to a point, so that by introducing it far up the canal greater strength can be obtained, and the root rendered less liable to longitudinal fracture from pressure in a forward direction. A piece of platinum plate, a trifle larger than the end of the root, of about No. 30 U. S. standard gauge, with a hole punched in its center, is then slipped on the post, which it must fit tightly (Fig. 140).

When the post is adjusted firmly in the canal, the platinum plate is pressed down on the root, and burnished into the orifice of the canal around the post. When the post is withdrawn from the root, the platinum will adhere to it, if fitted closely, without the use of wax. A particle of pure gold with borax is put in the joint, and melted in the flame of an alcohol lamp. Barely enough gold should be used to unite the parts. When soldered, the post and cap are again adjusted in the mouth and the cap malleted and burnished to the form of the end of the root, so that its edge will leave a mark on the platinum. The cap, on being removed, should be trimmed to this mark, and again burnished on the root (Fig. 141). Sometimes around the palatal portion of the root the platinum may be slightly burnished over the edge. The post is then cut off just above the platinum, and a plate tooth fitted, backed, and cemented with wax in position on the cap. The whole is then removed, invested, and soldered with gold at least twenty carats fine, which should be melted in at the base of the post, as at this point, when in use, the strain is very great. The post is then barbed, and the crown is cemented to the root with gutta-percha or oxyphosphate cement (Fig. 142).

Dr. F. T. Van Woert, of Brooklyn, N. Y., in constructing crowns of this style, shapes the end of the root, and adapts the cap as shown in Fig. 143. The slant given to the palatal side aids the root to resist force in a forward direction.

CHAPTER VI.

GOLD COLLAR CROWNS.

This style of gold crown includes those methods which consist in banding, capping, and hermetically inclosing with gold the end or the neck of a root, with or without any portion of a natural crown, for the purpose of securing stability to the artificial crown, preventing fracture of the root and decay of the parts, thus permanently preserving them. This method possesses much practical value as a preserver of tooth-structure and restorer of usefulness to the teeth, and affords excellent supports for bridge-work.

The use of the collar crowns was first described by Dr. Wm. H. Dwinelle in the application of the method to a crown with a porcelain front,[1] and by Drs. W. N. Morrison[2] and J. B. Beers[3] in the construction of all-gold cap crowns.

Collar crowns of which the part that essentially constitutes the cap is constructed in sections, will be first described.

THE CONSTRUCTION AND ADAPTATION OF COLLARS.

Careful study of the different forms of crowns and roots, and of the anatomical structure and relationship of the contiguous parts, is most essential for the perfect construction and adaptation of collars, bands, or ferrules, as they are variously designated.

Many devices and methods in use facilitate this operation, but its skillful performance can only be attained by study and practice, as is proved by the easy and perfect manner in which it is done by experts in crown- and bridge-work, who use no appliances but pliers and shears guided by an intuitive perception of the requirements of each case.

---

[1] *American Journal of Dental Science*, April, 1855.
[2] *Missouri Dental Journal*, May, 1869.
[3] Circular to dental profession, 1873.

The collar is preferably made of coin gold, or of 22 to 23-carat gold plate. Pure gold plate lined with platinum is also used, and iridio-platinum plate in special cases.

Gold plate of No. 32 to No. 34, or gold and platinum or iridio-platinum of No. 34 or No. 35, U. S. standard gauge, affords the requisite strength, together with easy adaptation to the form of the crown or root. The natural crown or root having previously been properly prepared (see page 39), a strip of the metal is cut of the length required, and generally from one-fourth to one-half of an inch in width (Fig. 144). The end to form the underlap is beveled with a file. The strip is then bent with suitable pliers (Fig. 145) to the average form (Fig. 146), any special deviation from such average being noted (Fig. 147), and to the size of the cervical periphery of the root of the tooth to be crowned. It is then placed on the root and adapted as closely as possible to its form, with the upper edge of the metal pressing gently under the free edge of any portion of the gum it may meet. It is then removed and cut so as to allow the ends to lap over slightly. The adaptation to the root is then continued, during which process the metal should be heated and chilled in water after each trial, in order to maintain the shape given to it. At the last adjustment to the root, the lap-over is marked on the metal with a sharp-pointed instrument. The joint is made at this mark by placing there the least possible quantity of solder and holding the collar in the flame of an alcohol lamp or a blue gas flame. The collar is then slipped on the point of a small anvil, and the joint tapped down and trimmed level.

Fig. 144.

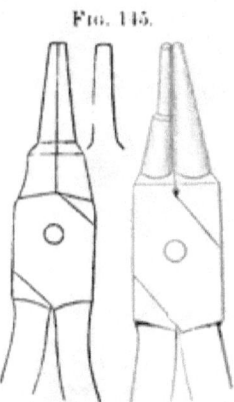

Fig. 145.

When a mandrel is used in forming a collar, the size or shape is first taken by encircling the root with a piece of fine iron or copper wire, about No. 28 U. S. standard gauge, and twisting the ends together on the labial side. The wire is then pressed up on the root about as far as the upper edge of the collar is to go, and burnished to the sides (Figs. 148 and 149). The wire ring is then carefully removed, laid on a piece of air-chamber tin, a piece of flat iron put over it, and with a blow from a hammer on the iron the wire is driven into the tin (Fig. 150). The wire ring is removed from the tin, slipped on a mandrel[1] that represents the form of the root to be crowned and pressed down gently as far as it will go without stretching the wire (A, Fig. 151). The distance from the end of the mandrel to the wire is then measured and marked on a strip of paper, and the wire removed. The gold to form the collar is then bent and shaped on the mandrel, with the edge which is to form the cervical portion (B) placed a little below the line of the wire (A), as shown by the measurement previously taken. The ends of the gold are beveled, slightly lapped, and the edge of the lap-over marked (C). The collar is then removed from the mandrel and, the ends being held together with common tweezers, which are grasped by pliers, or better still, by a small hand-vise (Fig. 152), the extreme

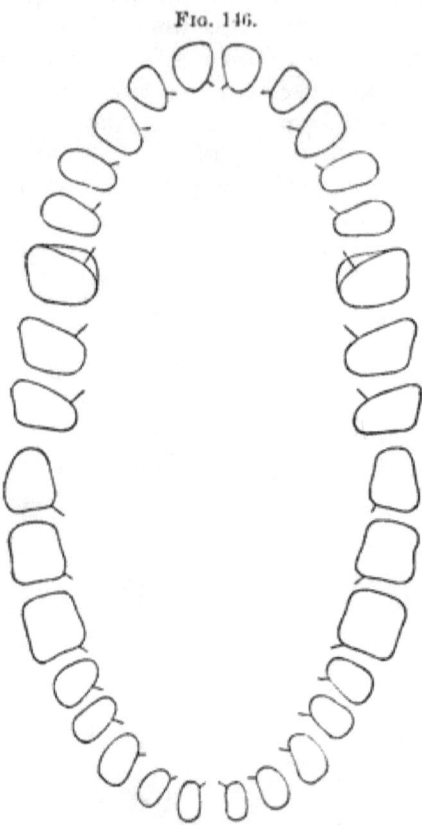

FIG. 146.

The palatal side of the superior molars, in many cases, is of the large oval form indicated by the outer line to the form of the first molar. The small spurs indicate the points generally found the most suitable to make the joint.

---

[1] A description of mandrels will be found in the chapter on the "Mandrel System."

outer end of the joint is united by an atom of solder with a blow-pipe. The points of the tweezers prevent the solder from flowing along the joint, the cervical portion of which is left open

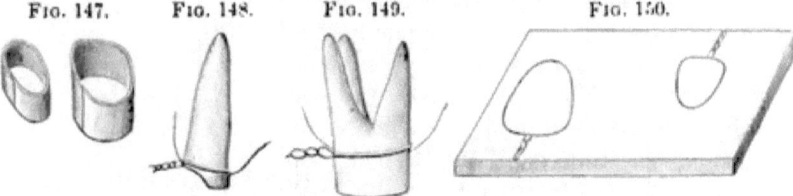

Fig. 147.   Fig. 148.   Fig. 149.   Fig. 150.

for the present. The collar is then shaped to the form given by the wire in the tin, after which it is ready for adjustment in the mouth. The unsoldered end of the joint permits the collar to be easily and accurately adapted to the root, after which the solder can be flowed across the collar and the joint closed.

When the collar has been formed, it is adjusted on the root and pressed or, by the aid of a piece of wood, one end of which

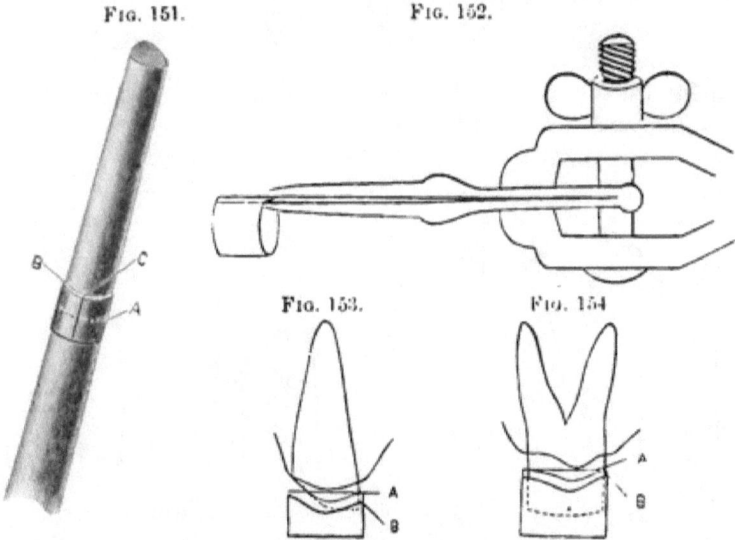

Fig. 151.   Fig. 152.

Fig. 153.   Fig. 154.

is placed across the outer edges of the collar, tapped, up to the margin of the gum. A line parallel with the margin is marked with a sharp-pointed instrument on the collar (A, Figs. 153 and

154), which is then removed, trimmed to this mark, readjusted, and again marked (B), and the process continued until the collar fits proportionately under the margin of the gum. If, on adjusting, the collar is found a trifle too small, it is easily enlarged by tapping the gold with a riveting hammer on an anvil as shown in Fig. 155. If, on the contrary, the collar should prove to be too large in circumference, the difficulty can be remedied by slitting the gold partly across the side opposite the joint, lapping the edges slightly, and soldering. The edge is then burnished to the periphery of the root. For this purpose a set of burnishers should be used especially formed to suit the different positions and avoid irritation of the margin of the gum. Such a set is illustrated by Fig. 156.

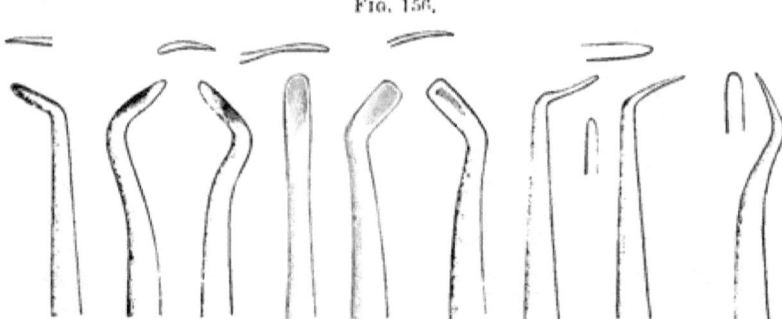

Fig. 155.

Fig. 156.

The application of local anesthetics such as cocaine, carbolic acid, or a mixture of tincture of aconite-root and chloroform, will lessen the pain attending the operation.

CHAPTER VII.

GOLD COLLAR CROWNS WITH PORCELAIN FRONTS.

INCISORS AND CUSPIDS.

This style of crown for incisors and cuspids, as originally made by Dr. C. M. Richmond, and with which his name has become associated, consisted of a cap for the root, formed of a band of gold capped with platinum on which was soldered a tooth with a slot in the center between the pins. Through this slot and the center of the cap a screw passed which entered into a cylinder previously screwed and cemented into the root-canal.

The form of gold collar crown in general use at present is, in principle, the same as what has been known in dentistry as a gold pivot tooth, with the addition of a gold collar for the root, and having the advantage of oxyphosphate for its cementation. These improvements, however, enhance its value as a crown, and materially change the process of its construction. In making an incisor or cuspid crown of this style, the collar, having been formed, is trimmed even with the surface of the end of the root. With the collar in position on the root, a corundum-wheel is passed over the labial edge, along the margin of the gum, to level the gold with the root and render it invisible when the crown is finished. The work will then present the appearance shown in Fig. 157.

The cap is made by adapting the surface of a very thin piece of platinum plate to the outer edge of the collar, and uniting them with solder in the flame of a lamp (Fig. 158). The quantity of solder used must be very small, and it should be placed on the platinum outside of the collar, as otherwise it will flow over the inside of the collar and interfere with the fit of the cap. The process is facilitated by first merely attaching the platinum, with the solder, to the edge of the collar, then readapting, and finishing the soldering. The platinum is then trimmed to the collar, and the cap adjusted on the root. The labial section of the surface of the cap is then burnished to the end of the root (A, Fig. 157). The root-canal having been slightly enlarged, a pin of round iridio-platinum wire, No. 16 or 17, U. S. standard gauge,—filed a little smaller for laterals or other roots which require it,—is slightly tapered at the point, fitted to an aperture made in the cap, and to the canal (B). The pin is then cut off even with the cap, removed, and temporarily laid aside.

A hollow wire, the open space in the center of which is very small, has lately been introduced for use in crown-work by Dr. J. G. Morey. The advantage it confers is the comparatively easy manner in which it can be drilled out of the canal if for

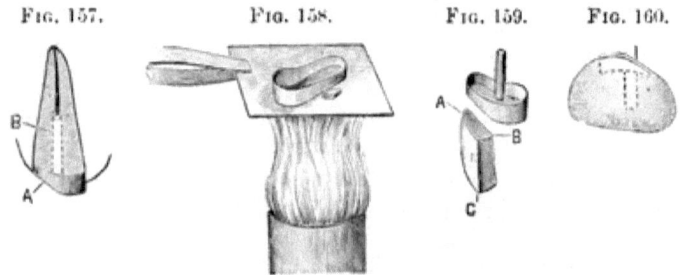

FIG. 157.   FIG. 158.   FIG. 159.   FIG. 160.

any reason it becomes necessary to remove the crown, as the drill will follow the fine opening in the center of the wire.

A plain-plate cross-pin tooth, suitable in form and color, is ground and fitted in position on the cap. The labio-cervical edge of the porcelain (A, Fig. 159) should be flush with the edge of the collar, and meet the margin of the gum. It should be cut out at the base (B) so as to form a slight space just over the end of the pin. The tooth is then backed with very thin pure gold, gold lined with platinum, or pure platinum. Platinum gives a faint blue shade, and gold, or gold lined with platinum, if the gold side is toward the porcelain, a slight yellow shade. The backing should extend as far as possible under and between the tooth and the cap, as the solder will flow in and fill the space, thus giving strength and continuity of structure. The backing, if bent over the incisive edge (C) at a right angle, will protect the porcelain in occlusion. A narrow strip of fine gold placed transversely across at that point previous to investing, and united in the soldering of the backing, will answer the same purpose.

The tooth, when backed, is secured in position on the cap with a compound of wax and resin, and the whole adjusted in the mouth, then removed, and the pin, which has been laid aside, warmed and placed in position by passing the end from the inside of the cap through the hole into the wax attaching the porcelain crown. Another adjustment in the mouth is then made to determine the exact line for the pin, and the case is ready for investment.

After the fitting of the pin to the root and root-canal, as has been described when the cap was formed, some prefer to solder it to the cap, for which purpose it should be adjusted in position and cemented with wax, then removed, invested, and soldered (Fig. 160). At this stage of the work, if desired, an impression can be taken in a small impression cup (Fig. 161) with either plaster or modeling compound. The cap should be removed in

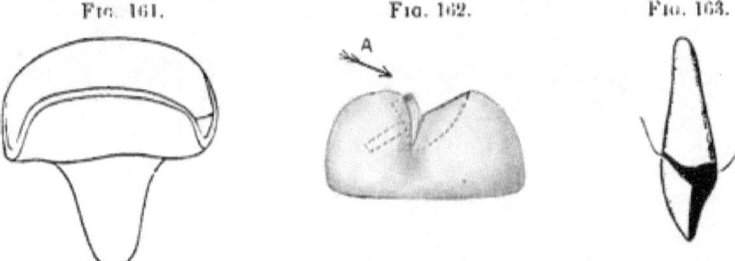

Fig. 161.   Fig. 162.   Fig. 163.

position in the impression. To aid this, a piece of gutta-percha can be put on the end of the pin projecting from the cap. An articulation of the lower teeth should also be obtained. When the model is made, the pin is cut off even with the cap, and the porcelain tooth fitted as already described.

Calcined marble-dust and plaster, in the proportion of two parts of marble-dust to one of plaster, to which is added a pinch of sulphate of potassium to quicken the setting, makes what is considered to be the most suitable investing material for crown-work. The crown, when invested, should be left exposed at the sides, as illustrated in Fig. 162. The flame should be pointed in the direction A, and the investment heated uniformly, so that the solder is flowed between the porcelain crown and the cap. Only sufficient gold should be used to insure restoration of contour. When soldered, after having been placed in acid and thoroughly divested of borax, the crown is ready for the finishing and polishing process. Fig. 163 represents the completed crown.

BICUSPIDS AND MOLARS.

Bicuspids crowned by this method will have greater strength if a portion of the palatal section of the natural crown is reserved (Fig. 22), and the band or collar made deep enough to cover it.

The end of the root is capped after the manner of the typical central already described, one or two pins being used in the canals as required. A porcelain cuspid tooth, or a bicuspid

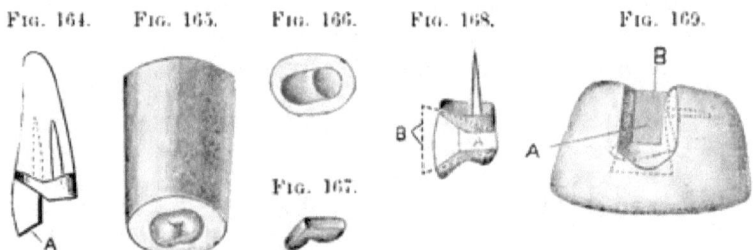

Fig. 164.   Fig. 165.   Fig. 166.   Fig. 168.   Fig. 169.

Fig. 167.

front, is then ground, backed, and adjusted on the cap to represent the labial aspect, and secured with wax. The front and cap are then removed, invested, and soldered, after which they are adjusted in the mouth, and the occluding edge of the porcelain is ground clear of the antagonizing teeth (A, Fig. 164). With a die of suitable size representing the occluding surface of a bicuspid, as illustrated in Fig. 165, a thin piece of pure gold plate is swaged (Fig. 166) and the cusps filled in with 18- or 20-carat gold plate. The cap is then trimmed (Fig. 167), ground, and fitted to the occluding edge of the porcelain front (Fig. 168) in proper position as regards occlusion, and the wax attaching it is shaped to the contour of the crown (A). A piece of pure gold plate (B), not over 34 or 35 U. S. standard gauge, is then adjusted on each side of the crown, which is invested (Fig. 169).

The long ends of the two side-pieces of gold plate are designed to retain them in position, as the investing material should be removed from the portion inclosing the sides of the crown (A). In the process of soldering the solder is placed in the aperture at B, and the flame of the blow-pipe being directed on the exposed sides of the gold at A, the solder is flowed into every part, forming perfect continuity of structure of the metallic portion of the crown. In finishing, the surplus gold is trimmed to the contour of a bicuspid tooth. Fig. 170 represents the finished crown.

The method described produces a perfect and artistically formed crown, but simpler and quicker methods are practiced. One of these is to build up the palatal cusp with several pieces of gold plate, which have been previously melted into the form of small balls and flattened out on an anvil. These, laid in position and united with solder, are shaped in finishing to represent

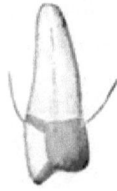
Fig. 170.

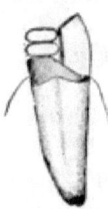
Fig. 171.

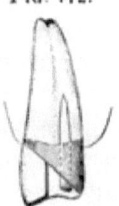
Fig. 172.

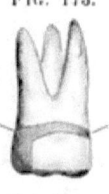
Fig. 173.

the palatal cusp (Fig. 171). The porcelain front should be backed so that the solder can be flowed over its occluding edge.

Another method is to extend the palatal part of the collar down as shown in Fig. 172, and then fill in the space with solder. In finishing, the gold is trimmed to the form of the crown. In this and in the method just previously described, the porcelain front can be soldered and the palatal cusp formed in the one investment.

The method of construction of molar crowns with porcelain fronts is similar to that for bicuspids (Fig. 173).

Dr. Stowell's Method.—A porcelain crown can be soldered on the cap, according to Dr. S. S. Stowell's method, as follows:[1] "The tooth used may be a Logan or an E. Parmly Brown crown

---

[1] *Dental Cosmos*, vol. xxix, page 641.

or an ordinary countersunk tooth, but in most cases the Logan crown is preferable. The pin is first cut off, then the tooth is ground to fit on the cap, porcelain and the stump of the pin being reduced alike evenly and smoothly; after which the stump of the pin is ground with a small wheel below the surface of the porcelain (Fig. 174). The tooth is then invested (Fig. 175) and pure gold fused on to the platinum pin, and while in a fluid state it is with a wax spatula 'spatted' down flat (Fig. 176). The gold is then filed or ground down even with the porcelain,

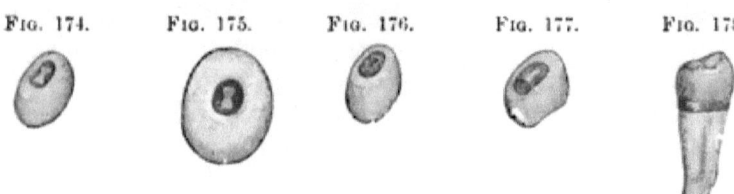

Fig. 174.   Fig. 175.   Fig. 176.   Fig. 177.   Fig. 178.

and at the palatal border the tooth is ground to bevel back until the gold is reached (Fig. 177). The tooth is then secured in place on the cap with wax cement (Fig. 178), the case invested, and heated until the wax has melted and burned out. A small clipping of thin platinum plate is crowded into the opening (Fig. 179) caused by the grinding of the bevel on the crown. The clipping of platinum serves as a lead for the solder, which follows it down into the countersunk cap, around the ends of the dowels, and finally attaches itself to the pure gold already attached to the stump of the platinum pin. Fig. 180 represents the completed crown. A sectional view of a like crown (Fig.

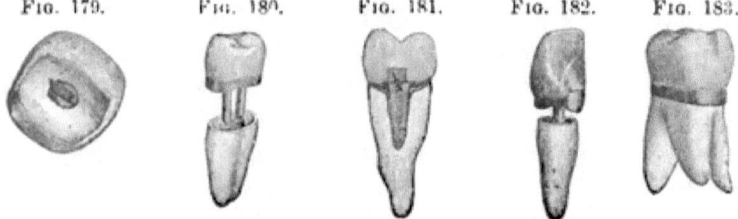

Fig. 179.   Fig. 180.   Fig. 181.   Fig. 182.   Fig. 183.

181) also shows the organization in detail. A porcelain crown can be used to represent any of the teeth in the same manner. See Figs. 182 and 183."

CHAPTER VIII.

ALL-GOLD COLLAR CROWNS FOR BICUSPIDS AND MOLARS CONSTRUCTED IN SECTIONS.

The root and crown having been properly prepared, the collar is formed and adjusted as described at page 85, and the edge toward the antagonizing teeth trimmed, to fully clear them in occlusion. The collar is then slightly expanded toward the occluding surface to effect contour, removed, filled with plaster, and adjusted in position. Fig. 184 represents a typical case. The antagonizing teeth, having been covered with a piece of tin foil, are then occluded until the plaster sets. The collar is then removed. The surface of the plaster inside the collar will give the impression of the natural root or crown, and the outside that of the antagonizing teeth. The latter furnishes an outline of the grinding-surface of the crown.

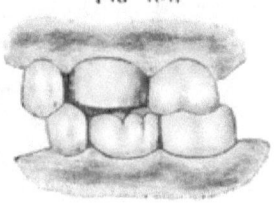

Fig. 184.

The plaster is then trimmed and shaped to represent the cusps and fissures of the natural tooth, enough of the surface being removed to allow for the thickness of the plate that forms the cap. Readjustment in the mouth will show the accuracy of the occlusion.

A small tube of copper, a trifle larger in circumference than the crown under construction, is filled with Melotte's "moldine," and the surface rubbed with soapstone. An impression of the lower portion of the form of the crown A to the line B, Fig. 185, is then made in the moldine, and a strip of paper wound around the tube, extending about an inch above the edge. Fusible alloy is then melted and poured into the mold, thus forming a die.

An indentation is made with a punch in a block of lead, into which the die, when cold, is hammered slightly beyond the impression of the edge of the collar. By this method a die and a counter-die (Fig. 186) can, with practice, be completed in five minutes. With this die the cap is then struck up on the lead from a flat piece of plate and fitted to the collar. A little of the surface of the plaster in the collar may have to be removed, if, on trial in the mouth, the cap is found a little flush. The crown, with the plaster still inside the collar, is fixed in a soldering clamp constructed in one of the forms shown in Figs. 187 and 188, which holds the parts together and permits the flame to reach all points. No more solder should be used than the contour requires, as an excess necessitates additional labor in finishing.

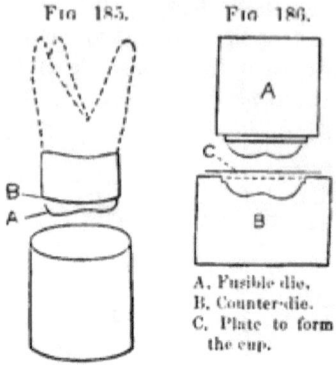

Fig. 185. Fig. 186.

A. Fusible die.
B. Counter-die.
C. Plate to form the cup.

Another method, if the crown is not to be contoured with the aid of the solder, is, when the cap is struck up, to melt solder into the cusps, and then adjust the cap in position on the collar, for which purpose some of the plaster underneath the cap must be removed. A jet of flame from the blow-pipe is then thrown

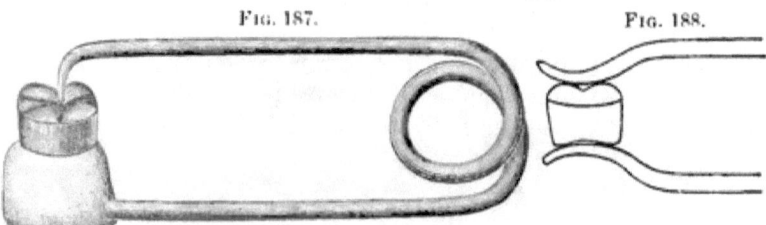

Fig. 187. Fig. 188.

The frame of this clamp is formed of iron wire, and the support for the crown of plaster, asbestos, and marble-dust.

upon it in such a way as to cause the solder to flow down on the edge of the collar and fill the seam from the inside. The objection to this method is that, when a large portion of the natural crown is inclosed by the gold, the solder will occasionally alter the inside form of the fitted cap, thereby interfering with its adjustment, which is a defect troublesome to correct.

Still another method is to adjust the collar in the mouth, and, with a small piece of wax or impression-compound pressed upon it, to take an impression and "bite," in which the collar will be imbedded and removed. With this a model and articulation are made and the form of the cap shaped in wax. An impression of the cap is then made, either in moldine in a soft state in a tube, or in plaster, and a die cast. The cap is stamped on this die, then adapted to the collar by the model, and the crown finished. This method, which was first made known by Dr. N. W. Kingsley, is adopted when it is preferable to construct the crown between the visits of the patient.

In utilizing a tooth as an abutment in bridge-work when all or nearly all of the occluding surface of the natural crown is present, a practical method of construction is to mark the outline of the natural crown on the inner surface of the collar; then remove the collar and trim so as to leave a border of about one-sixteenth of an inch outside the mark. This border is then

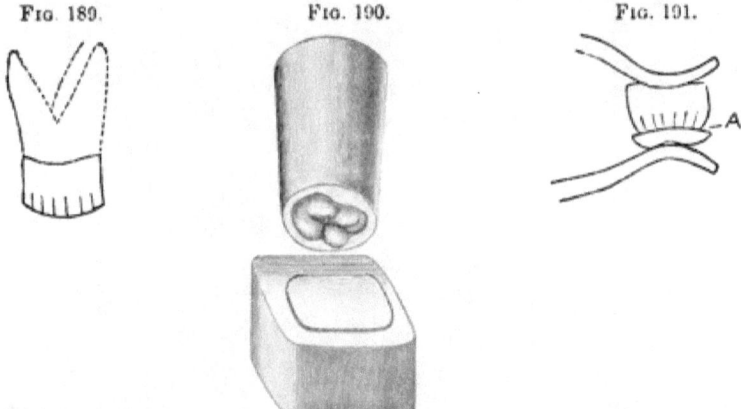

Fig. 189.   Fig. 190.   Fig. 191.

thinned with a corundum-wheel, and slit as seen in Fig. 189. The collar is next adjusted on the crown, and the slit border bent over to the form of the occluding surface, to which it is burnished. A piece of pure gold plate, about No. 34 gauge, is then placed on the occluding surface of the tooth and adapted to it and to the collar. The gold may first be struck in the form of a cap by laying it on a block of lead and hammering into it a die corresponding to the surface of the tooth to be crowned

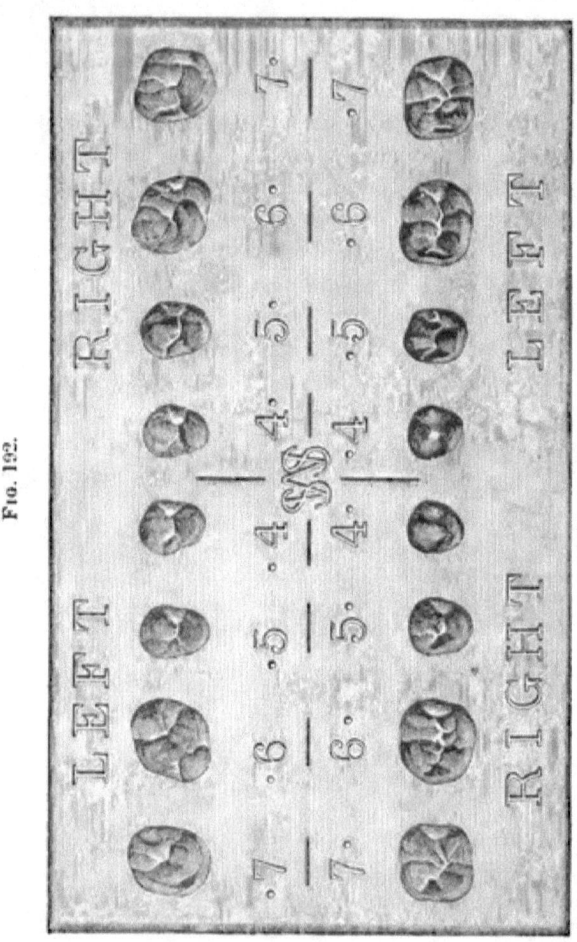

Fig. 192.

(Fig. 190). The antagonizing teeth are then occluded on the gold, which is thereby pressed to form to articulate with the occluding surfaces. Enough of the occluding surface of the tooth crowned should always be removed to allow for the thickness of the gold covering its surface. The collar and cap are next removed and soldered. This is done by resting the collar on the cap, which is held by a pair of tweezers, or by clamping the cap and collar together and placing the solder in small pieces around the collar outside the cap, at A, Fig. 191, and soldering by holding in a blue gas flame. Only sufficient solder should be used to join and fill the seams, so that it will not interfere with adjustment on the natural crown.

The methods described insure a perfect occlusion of the crown with the antagonizing teeth. In the absence of antagonizing teeth, or when the general form of the grinding-surface permits

Fig. 193.

it, the cap can be struck up with a die similar to the one shown in Fig. 190. The cusps are then filled in, and the edges of the inner surface of the cap ground level on the side of a corundum-wheel. The entire circumference of the edge of the collar is also leveled, and the cap adjusted, clamped, and soldered. If the cusps of the cap are filled in with solder, it will flow down and join the collar on the inside; if with gold plate, the cap and collar must be joined with solder either on the inside or outside.

Metallic caps, or forms of the occluding surfaces of teeth for use in constructing crowns, are quickly made with the die-plate shown in Fig. 192,[1] "in which are four groups of intaglio dies representing, with distinctive correctness, the peculiar cusps of

---

[1] *Dental Cosmos*, vol. xxix, page 482.

the upper and lower right and left bicuspids and molars. These are indicated by the Hillischer notation, so that each form may be easily identified in practice. The hubs A, B (Fig. 193) are of the sizes shown, and are made of an alloy composed of tin one part, lead four parts, melted together. The mold C should be warmed, the metal alloy poured in every hole, and the overflow wiped off just before the metal stiffens; this will make the butts of the hubs smooth and flat. After a minute or two the mold may be reversed, the hubs shaken out, and the casting process continued until a considerable number of hubs shall have been made. In Fig. 194 a molar hub is shown in place on a piece of No. 32 gold plate, which lies over the 6· (upper right first molar) die. A succession of blows on the hub with a four-pound smooth-face hammer will drive the plate into the die, and, at the same time, spread the hub metal from the die center to its circumference, in

FIG. 194.

  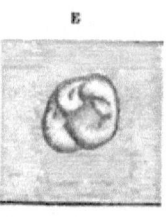

FIG. 195.

such a manner that the plate will be perfectly struck up with the least possible risk of being cracked. The flattened hub is seen in Fig. 195, which also shows at D the obverse of the struck-up hub, and at E the cameo of the struck-up plate, having every cusp and depression of 6· sharply defined. The counter-die plate

ALL-GOLD COLLAR CROWNS CONSTRUCTED IN SECTIONS. 101

(Fig. 192) is made of a very hard cast metal, which will admit of the striking up of many crown plates by the means described, if the crown plates be not too thick and stiff. Of course they should be annealed before they are placed over the die.

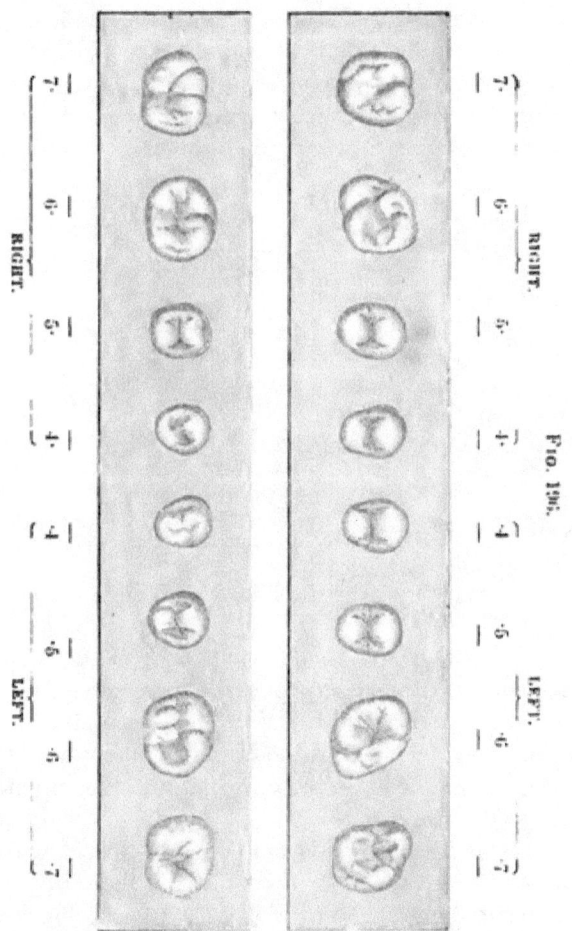

FIG. 195.

"In careful hands the die-plate should give clear cusp definitions after years of use. The counter-die plate is in some respects similar to a stereotype plate for printing, and the struck impressions on two strips of thin plate will therefore appear as in Fig. 196, wherein their regular order is noticeable, as seen from the cameo surface of the struck plates.

"The peculiar action of the hub in forming first the center of the crown plate, and spreading from the center outwards, as the hub is shortened under the hammer, until the die is overspread by the plate and hub, with the result shown in Fig. 195, is an essential feature of this process for obtaining easily and quickly the superior styles of coronal cameos shown. If a cusp or fissure should chance to crack in hubbing, a small piece of

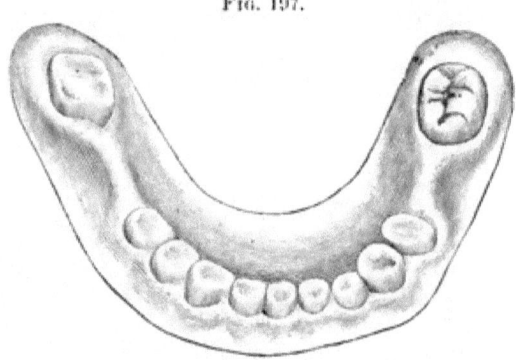

Fig. 197.

plate may be struck up over the fissure, and then soldered to the original cap."

The methods which have been described for the construction of all-gold bicuspid and molar crowns are those generally adopted in practice. Of others, Dr. J. J. R. Patrick's method[1] consists of first forming a very narrow collar and telescoping it with a seamless cap of the form of the crown, and soldering along the line of the cap to the collar.

Dr. E. P. Brown's method is to make or select a metallic die for the crown to be formed; then place a piece of pure gold plate, about No. 31 gauge, on the flat surface of a block of lead, and gradually stamp the die downward into the gold to about

---

[1] Dr. Patrick's crown-work methods and the principles upon which they are based are set forth in a paper published in the *Dental Cosmos* for October, 1888, page 706.

half the depth of the intended crown. The gold is then removed, and each side of the unswaged portion slit and adapted to the form of the die, the full length of the crown, which is then adjusted in the mouth, the edges of the gold trimmed to the proper form, and the slits soldered.

Dr. M. Rynear's crown is of the same general character and is constructed in the same manner as Dr. Brown's, except that a seamless cap is used to form the crown, instead of the flat piece of plate used by Dr. Brown.

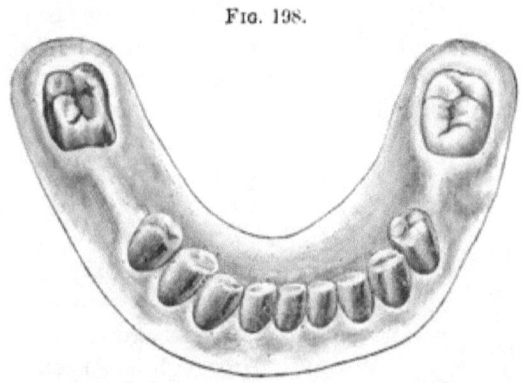

FIG. 198.

Fig. 197 represents a case of abrasion of the lower teeth, to which all-gold crowns have been applied, as shown in Fig. 198. In such cases, owing to the aggression of the occluding teeth, the cap forming the occluding surface should be constructed of heavy gold and platinum plate. An artificial plate replaces the upper teeth.

CHAPTER IX.

THE GOLD SEAMLESS CAP CROWN.

This method consists in the use of a gold seamless cap for the construction of the required root cap or crown.

INCISORS, CUSPIDS, AND BICUSPIDS, WITH PORCELAIN FRONTS.

Incisor, cuspid, and bicuspid crowns with porcelain fronts are constructed by this method as follows: The natural crown is ground down to within about one-eighth of an inch of the gum at the palatal wall, or enough to clear the antagonizing teeth when occluded, and slanting from the posterior edge of the pulp-chamber to the cervico-labial edge of the gum and slightly under its margin if it is desirable to conceal the joining of the crown with the root. The sides are shaped the same as for a collar crown (Fig. 199). A die of the end of the root is then made. For this purpose an impression of the part is taken with

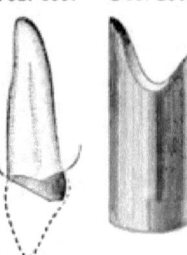

Fig. 199. Fig. 200.

gutta-percha on the end of a piece of wood trimmed to the proper size, or by placing the gutta-percha in a tube formed of a strip of copper about one and one-half inches in length and three-eighths of an inch in diameter, cut out on the sides to the depth of half an inch, with the flange for the palatal side shortened (Fig. 200). The impression thus taken will be confined almost entirely to the end of the root to be capped. When the gutta-percha has cooled, a strip of paper is tied around the wood or tube and a die cast with the fusible metal. When cool, the die is removed from the mold, and the metal is trimmed, with file and chisel, a little deeper than the gum has permitted the impression of the root to be taken, and without altering the form of the end of the root (Fig. 201). A counter-die is then made by punching a hole in the surface of a block of pure lead, and with a few blows of a hammer driving the die into it.

A cap of gold can be formed by placing a piece of gold plate (preferably pure, No. 32 U. S. standard gauge) of suitable size upon a block of lead, and with an oval-shaped punch one-fourth of an inch in diameter gradually driving it into the lead until the gold has assumed the shape of a cap about a quarter-inch in depth (A, Fig. 201). The gold should be withdrawn from under the punch and annealed several times during the process. Caps can also be made with a stamping-press such as was introduced by Dr. J. J. R. Patrick, of Belleville, Ill. (See page 112.)

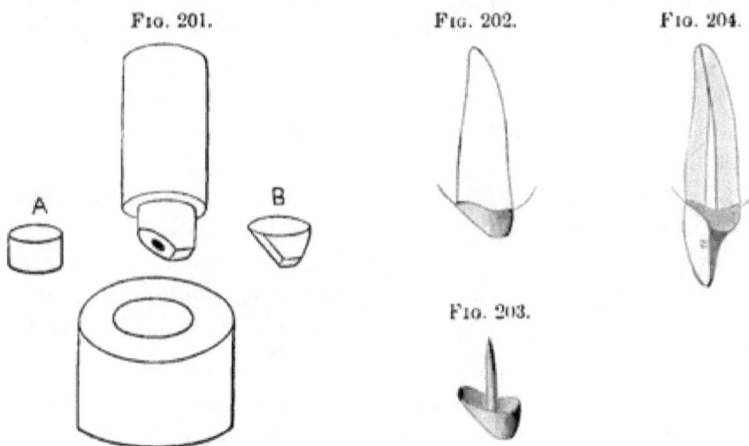

The cap is then annealed and swaged on the die to the form of the end of the root (B, Fig. 201). The palatal portion of the cap should be allowed to go well up under the free edge of the gum, and at the cervico-labial edge it can be, if preferred, cut out to the edge of the root. In the process of adjustment, the edges which fit under the gum should be marked and trimmed as directed in describing the construction of a collar crown, and then burnished close to the sides of the root and into the orifice of the root-canal, forming a perfect-fitting seamless cap (Fig. 202). An iridio-platinum pin is then fitted in the root-canal and soldered to the cap (Fig. 203), or afterward adjusted as in the construction of the gold collar crown with porcelain front (page 90), with which operation the remainder of the process of construction is identical. Fig. 204 represents the completed crown.

The advantages of this style of crown are, simplicity, as the formation of a collar is avoided, and strength, as a large portion of the natural crown can be left at the palatal side. This affords a stronger and more reliable foundation than can be obtained at any other point, as the direction of the force in mastication is forward at an angle with the line of the root, and although the gold of the cap, where it encircles the root at the cervico-labial edge, is entirely removed, the crown is still held securely.

In a paper on the subject of preserving and utilizing this part of the tooth, Dr. W. F. Litch, of Philadelphia, describes a crowning operation.[1] He constructed the cap of platinum by slitting a piece of the plate in a number of places, adapting it to the form of the end of the root, and then soldering the whole together.

This operation is not, however, so easily or so satisfactorily performed as the method above described, in which platinum, if desired, can be used instead of gold, and the soldering done with 22- or 24-carat gold. In some respects pure platinum is preferable to gold in capping roots, as it is less likely to be affected by the secretions of the mouth.

ALL-GOLD BICUSPIDS AND MOLARS.

All-gold seamless crowns for bicuspids and molars that will accurately fit the natural crown and root, and occlude properly with the antagonizing teeth, are easily and quickly formed, if sufficient of the natural crown remains to admit of temporary restoration of its contour with gutta-percha or any other suitable plastic material. When this has been done, an impression of the restored tooth is taken in gutta-percha in a tube, as explained on page 104, and a die then formed of fusible alloy; or a plaster model can be made from an impression of the tooth taken in wax, and a mold obtained from the model.

The preparation and shaping of the natural crown to receive

---

[1] *Dental Cosmos*, vol. xxv. No. 9, page 449.

the artificial crown can then be proceeded with. Where the natural crown is very badly decayed or broken down and the method just described is not practicable, the portion of the natural crown or root remaining should be shaped and prepared to receive an artificial crown. Then the form of the cervix is ascertained with a wire as described on page 86, and an impression of the parts taken in wax, and the wire form, the twisted ends having been shortened, is carefully adjusted on the wax at the cervical line. The plaster model, when made, will show the wire slightly imbedded in the plaster. The plaster should be trimmed to the inner edge of the wire, as that represents the exact form of the root (Fig. 205).

Another method is to encircle and adapt to the form of the root a strip of annealed copper, No. 35 standard gauge, about

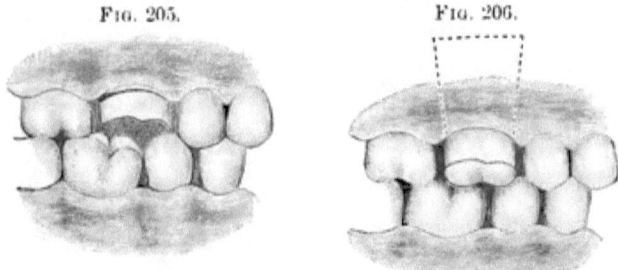

Fig. 205.   Fig. 206.

one-eighth of an inch wide. The copper is then removed, the ends heated and touched with resin and wax, the copper adjusted on the root, and the ends cemented with a warm instrument. An impression is next taken with plaster, in which the copper band is removed in position. A model made from this impression, after the copper is removed, presents the exact form of the root.

From a "bite" taken in wax a plaster articulation is then made to the model. A hole is then drilled in the center of the form of the root on the model to be crowned. In this hole, and over the end of the root, a ball of soft plaster, slightly colored with carmine, is placed, and the teeth of the articulation, covered with tin foil, closed on it. This, on separation, gives the outline of the form of the grinding-surface for the crown. The sides of the plaster are then trimmed to the form of the crown, and the whole carved in detail (Fig. 206). As the crown will always stamp larger in circumference than the die, in proportion to the thickness of the gold used, an allowance must be made by trimming off a proportionate amount of the surface of the model; otherwise the outer surface of the crown will be larger than is desired. When the plaster model for the crown is made, it is separated from the rest of the model at the dotted line seen in Fig. 206 and trimmed in the form shown by the cast A, Fig. 207. From this model the die is made in a tube with moldine and fusible metal as described at page 104. The cast should always be lengthened at the neck, so that the crown when constructed shall have a surplus in depth of gold to allow for any trimming or shaping of the collar that may be required. The counter-die (B, Fig. 207) is made by punching a hole in

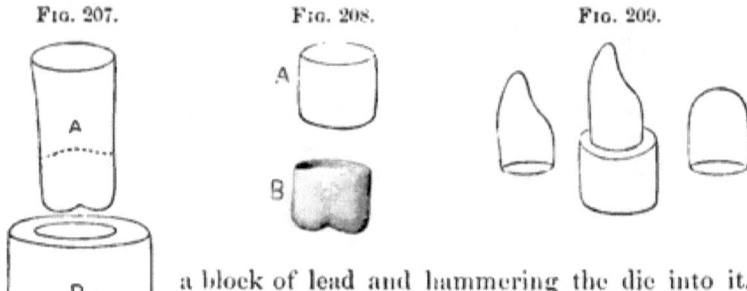

Fig. 207.   Fig. 208.   Fig. 209.

a block of lead and hammering the die into it. The crown, which is usually formed of pure gold, or gold slightly alloyed, or gold lined with very thin platinum from No. 32 to No. 34 U. S. standard gauge, is then made by first stamping a piece of plate (see page 110) in the form of a cap of gold (A, Fig. 208). This cap is then placed on the cast and with the aid of the counter-die (B, Fig. 207) swaged to the form of the crown B, Fig. 208. A piece of kid leather should be used to cover and protect the gold from the lead, and facilitate its removal from the counter-die. An allowance for the thickness of the leather must be first made, by driving it, without the gold on the cast, into the counter-die to enlarge it. If this is not done, the gold is liable to be torn in the swaging.

Cuspid crowns from which a portion of the gold on the labial aspect is to be removed, or which are to be used entire as a support for bridge-work, can usually be advantageously formed with a seamless cap (Fig. 209). The necks of these crowns can be contracted in fitting in a contracting plate, or slit, lapped, and soldered, should the case so require. (For details of process of adjustment and insertion, see " Process of Adjustment of Seamless Contour Crowns.")

CHAPTER X.

GOLD SEAMLESS CONTOUR CROWNS.

The artistic requirement of all-gold crown-work is, that it shall reproduce the anatomical contour of the natural teeth. This is usually accomplished by melting solder on the collar and then trimming it to the form of the crown. A preferable method is to shape the metal forming the sides of the crown by swaging. This is easily done in a crown formed in sections, but a special process is required in the construction of seamless crowns.

A contour crown can be made by placing a seamless cap on a sectional die or mandrel of the shape of the tooth, first swaging

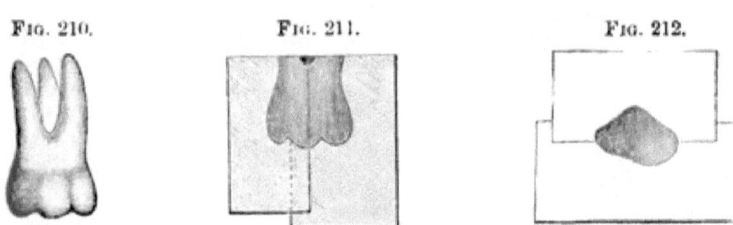

Fig. 210.   Fig. 211.   Fig. 212.

the grinding-surface on the mandrel and then stamping downward on the straight sides of the crown with a cap fitted to the shank part of the mandrel. But such a process, like many others, is too complicated to be of any use to the dental practitioner. The sectional mold method here presented is simple, practical, and general in its application.

To describe and illustrate the process, we will take one of the most difficult crowns to construct,—a superior molar (Fig. 210). A natural tooth, or one made of plaster, is used as a model. From this a sectional mold is made, as illustrated in Figs. 211 and 212, in Babbitt's metal, zinc, or fusible alloy.

Into the mold a cap of gold (Fig. 213) 23 to 24 carats fine, 30 to 32 gauge, is adjusted, fitting tightly the orifice of the closed mold. The mold is placed in a vise, the cap expanded to the general form of the mold by hammering into it a mass of cotton, and then swaged more in detail to the form, and with a wood point or a burnisher revolved by the dental engine burnished into every part of the mold (Fig. 214). To facilitate the process, the mold should be frequently opened, and the gold annealed. Fig. 215 represents the completed crown. These results can be secured by other styles of molds: Fig. 214 illustrates one, but the principle is the same.

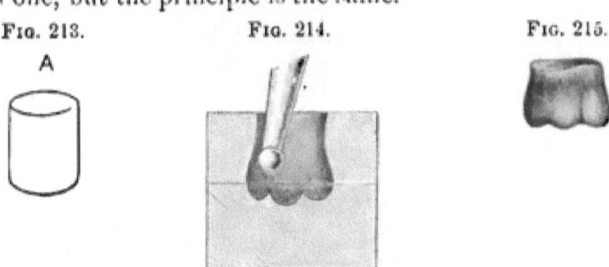

FIG. 213.   FIG. 214.   FIG. 215.

Another method is to form a fusible-metal die of the tooth to be crowned, and, after having stamped the grinding-surface of the crown, to reverse and swage the sides close to the die; the crown is then relieved of the core (die) by heating to the melting point of the fusible metal and pouring it out.

For practical use, a variety of molds should be made from natural teeth of different sizes and average forms to serve in corresponding cases. The crowns should be contracted at the neck more than their size and contour call for, so that the gold will act as a tight-fitting band which will expand to the form of the root as the crown is pressed up in the process of adjustment.

Caps of metal can be made in different sizes and kept on hand for use in this and other styles of crown-work by means of a machine (Fig. 216), which in principle is such as is used by jewelers for forming cap-shaped pieces of gold, and in factories for making copper cartridges. The gold plate, cut into circular pieces, is pressed through a steel die-plate, with punches gauged to the holes; at each punch a small portion of the gold is turned over, thus preventing its lapping or creasing (Fig. 217). Repeated annealing of the metal is very necessary in this process.

Fig. 216.

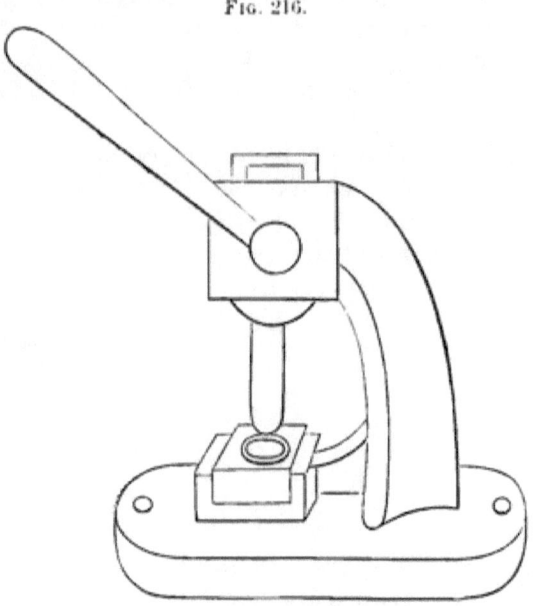

Methods of Contouring Crowns constructed in Sections.—In constructing a crown in sections, the collar can be first formed on a

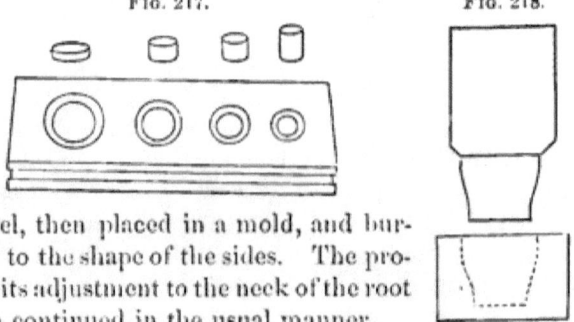

Fig. 217. Fig. 218.

mandrel, then placed in a mold, and burnished to the shape of the sides. The process of its adjustment to the neck of the root is then continued in the usual manner.

Another method is to stamp or burnish up the collar on a die representing the upper sections of a tooth, designated as the middle and cervical third (Fig. 218). After contouring the collar, the cap is adjusted and soldered on. With a metallic stamping plate (see page 98) these caps are quickly made.

Process of Adjustment of Seamless Contour Crowns.—A superior molar—one of the most difficult teeth to operate on—will serve as a typical case to illustrate this process. The crown or root is first shaped and if necessary built down with amalgam, straight, or tapering slightly on its sides toward the occluding surface, as described at page 41. Guided by the shape of the natural

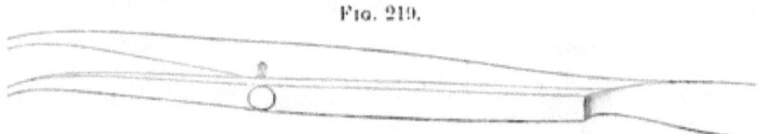

Fig. 219.

teeth and measurements by calipers,—a pair of tweezers with a screw as shown in Fig. 219 answers the purpose,—or by a plaster model of the mouth, a gold crown of suitable form and size is selected. The crown is then slipped over the end of the prepared natural crown or root, and gently pressed and worked straight upward, the gold at the neck of the crown expanding to the form of the root until the edge meets the margin of the gum (A, Fig. 220). A line (B) is then marked on the gold with a sharp-pointed instrument, parallel to the margin of the gum. The crown is then removed and the edge trimmed off to this mark

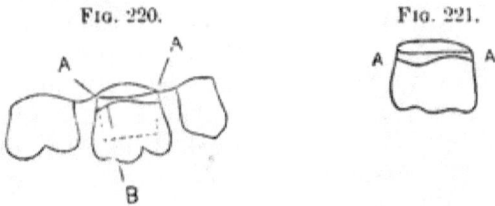

Fig. 220. Fig. 221.

or even with it (A, Fig. 221), using small curved scissors. This process should be repeated, taking off a little at a time, until the edge of the gold meets the gum evenly at all points, under the free edge of which it is then pressed up, and if the occlusion is correct a burnisher is passed around the cervical portion and the gold burnished accurately to the root.

Although a crown can usually be expanded on the root to its

8

form in the process of adjustment, the necks of natural teeth vary so in proportion to the size of the crown that it often becomes necessary to slightly expand the gold at the neck of a contour crown. This is quickly and easily done, without altering or injuring the crown, as follows: Soften a mass of gutta-percha, about the size of the crown, upon the closed ends of a pair of clamp forceps, or an appliance of similar construction. Introduce the gutta-percha inside of the neck of the crown, which should be moistened to prevent its adhesion. Then withdraw the gutta-percha, harden it in cold water, and cut through the center, between the points of the forceps. The points of the forceps armed with the gutta-percha are in effect an expanding sectional mandrel, and by reinserting them the neck of the crown may be expanded in any direction according to the manner in which the forceps and gutta-percha have been inserted (Fig. 222). A gold collar can be expanded in a similar manner.

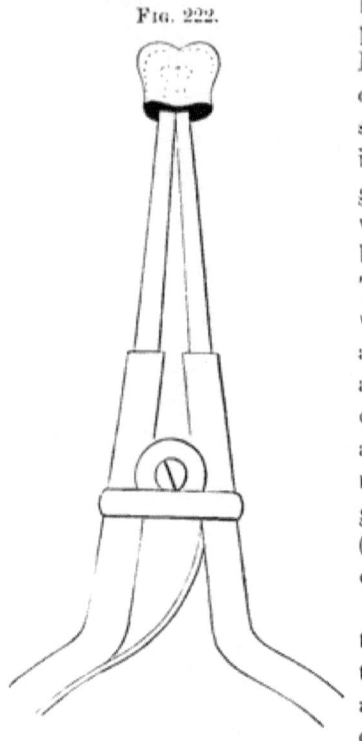

Fig. 222.

Before the crown is pressed up to its apparently proper position, the occlusion should be examined and calculations carefully made to obviate any defects, which at this stage are readily corrected by proper manipulation. Any desirable change can usually be made in the form of the occluding surface of the crown while on the tooth, with an instrument tapped by the mallet, or by removing the crown, holding it between the thumb and fore-finger with the second finger pressed flat against the edge of the cervical portion, and tapping the gold with a riveting hammer. The position can be changed for altering the sides, on which the flat face of the hammer should be used.

Strengthening Seamless Gold Contour Crowns.—Additional strength and stiffness can be given to seamless gold crowns when desired, in several ways. The liability of melting the gold which forms the sides of the crown in the operation has, with some, been the principal objection to their use. This, however, can be avoided. Where the cusps or grinding-surfaces require filling in or thickening, which is done from the inside, it is safely and easily accomplished by the proper use of prepared gold solder filings (see chapter on Plates and Solders, Part IV). The solder in this form, in the dry state, is carried with a spoon-shaped excavator, and packed in position in the cusps, or placed on any desired spot. The crown is then held in the flame of a small alcohol lamp and heated sufficiently to fuse the solder, which will melt down exactly where it is put, and not flow over the adjacent surface. During the process the crown should be held by grasping the edge of one side of the collar with the points of a pair of tweezers, in such a position that a full view of the inside surface shall be presented, and the melting of the solder thus determined instantly. Should the sides of the crown require strengthening, a small portion of the prepared solder filings can be moistened, and with a small camel's-hair artist's brush painted over the inside surface. When sufficient heat is applied, the solder will be fused evenly over the gold without the slightest risk of melting the sides or changing the general form of the interior of the crown. When both the cusps and the sides of the crown are to be strengthened, the sides should be done first. Another method is: Take a pellet of moldine, moisten it with water, and work it until quite soft, then roll it out lengthwise, and envelop the crown on its sides with a very thin continuous piece, leaving exposed the parts of the occluding surface that are to be filled in. Pass around on the outside of the moldine, without touching the crown, a fine wire, and twist the ends together for a handle (Fig. 223). Grind some borax mixed with water to a cream-like consistence, and with a small pellet of cotton twisted on an instrument paint the inside of the crown with the borax just where you wish the solder to flow. Then place inside the cusps an easy-flowing solder that has been cut very fine and immersed in the borax. Hold the crown in the flame of an alcohol lamp or in a blue gas flame, and heat slowly so that the position of the solder shall not be changed by the fusing of the borax. Increase the heat until the solder flows over the surface of the gold, which it does very quickly when the melting-point is reached. Then the crown must be instantly removed and examined. Sufficient solder—the filings previously referred to are safer for this purpose than solder cut in pieces—should be placed in the crown to give it the desired strength in one heating, but in case of necessity more can be added and the process cautiously repeated. The moldine protects the sides, which generally take up the greatest degree of heat.

Gold seamless crowns can also be strengthened or filled with solder or even 18- or 20-carat gold plate, by investing the outside

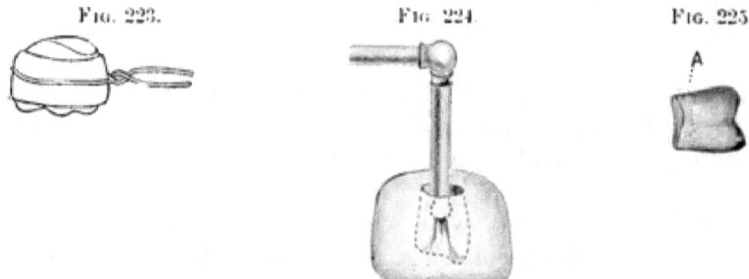

Fig. 223. Fig. 224. Fig. 225.

surface of the crown in plaster and marble-dust (Fig. 224), and then with a small flame of the carbo-oxyhydrogen blow-pipe, not over one-half an inch in length, introduced inside of the crown, melt and flow the solder or gold plate over any portion or even all of the surface of the gold. The crown, if formed of gold with a thin lining of platinum, can be soldered by either method with little danger of being melted.

If for some special purpose it is desirable to strengthen the neck of a crown by increasing the thickness of the gold on the external surface, the crown is first soldered to stiffen the occluding surface as above described. The inside of the crown is then filled solid with moldine or investing material. The gold is scraped around the crown, along the line of the cervical edge (A, Fig. 225), only as low as the stiffening is required. Along this part borax is applied with a brush, and small pieces of

20-carat solder placed and melted in succession, the crown being turned during the process until the neck is entirely encircled. A very fine platinum wire or a narrow strip of platinum foil adapted around on the outside of the crown, at the point soldered, lessens the quantity of solder required and facilitates the operation.

Supporting the Crown.—In crowning teeth with living pulps there is sufficient of the natural crown present to afford a secure foundation and attachment for the artificial crown, as is also the case with many teeth that are pulpless; but in badly broken-down crowns, or where only the root is present, a metallic pin or post should be inserted in the root, and the part built down with amalgam to a form which will afford secure support and attachment to the artificial crown, and facilitate its adjustment. (See "Special Preparation of Badly-Decayed Teeth or Roots," page 42.)

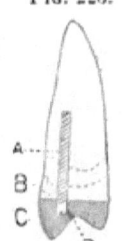

Fig. 226.

In many cases the required support for the crown can be secured by means of a screw (Fig. 226). A How screw is inserted in the root-canal A (see page 55). Amalgam is then packed in the lower section of the artificial crown, C, to the line B, and into the amalgam the screw is pressed. Amalgam which has been put in a piece of chamois and the mercury pressed out with a pair of pliers until it is in the condition termed "dry" will adhere to the gold without affecting it. The amalgam is first placed in the crown slightly in excess of the amount required, and the crown adjusted, removed, and the surplus scraped out. This process is continued until the screw or the crown section of the natural tooth forms an indentation in the amalgam, which it will fit when the crown is cemented on. The vent for the escape of air and surplus cement—which should always be put in perfect-fitting crowns and afterwards filled with gold or amalgam—should be in the line of the indentation in the amalgam, with which it must connect (D). (For process of cementation, see article on "Insertion and Cementation.")

These crowns can be inserted in an easy and inexpensive manner by filling in the lower section of the crown with amalgam instead of gold, and allowing the head of the screw or the natural crown to indent the amalgam as above described, and then cementing on the crown with oxyphosphate in the usual manner.

In a case so inserted, with no antagonizing teeth, the result is the same as though the inside of the occluding surface of the crown was filled with gold; but if antagonizing teeth are present, the gold of the crown is apt to wear through in places and expose the amalgam.

If a tooth is short and the articulation such as to require the reduction of the collar to a size which will suggest insecurity when the crown is cemented, or if the tooth is pulpless, a headed or barbed pin that will anchor in the root should be soldered in the interior of the crown, as shown in Fig. 227. This is done by passing the pin through a hole made in the occluding surface of the crown, adjusting the crown in the mouth, removing, investing, and soldering the pin from the outside surface of the crown.

Fig. 227.

The advantages of seamless contour crowns are, that they represent perfectly the tooth in its anatomical contour, present a uniform surface of pure gold, which preserves its color without tarnishing, and are quickly and easily adjusted. Their defects are inability to meet the requirements of abnormally-shaped roots and anomalous articulations.

CHAPTER XI.

GOLD CROWNS WITH PORCELAIN FRONTS FOR TEETH WITH LIVING PULPS—COLLAR CROWNS HYGIENICALLY CONSIDERED.

In the anterior teeth, in case of atrophy or erosion, or where decay has destroyed the approximal sides of a tooth in such a manner that crowning is considered the most desirable operation to perform, the pulp is frequently found unexposed and in a normal condition. The importance of its preservation in such a case is unquestionable.

The methods at present most commonly used to form a crown under such conditions are in many respects defective and objectionable, notwithstanding the advantages their indorsers claim for them.

In the first place, all crowns of platinum or iridio-platinum with body baked on to represent the labial aspect of the tooth, have a dead and unnatural appearance, and from those on which films of porcelain representing teeth are baked the porcelain frequently chips off, and both styles usually protrude beyond the line of the adjoining teeth.

The method here presented is intended to overcome these objections and to produce a more satisfactory result. The descriptive details of the crowning of a central incisor will serve to illustrate it. A model is first made representing the tooth in a perfect form. This can be done either by shaping the natural tooth in the mouth with oxyphosphate or gutta-percha, taking an impression of it in wax or moldine, and forming a model in plaster or fusible alloy, or by taking a natural tooth and shaping it to correspond, or, if the operator is an expert, carving one from a piece of plaster.

Two casts, one of the coronal form of the tooth (Fig. 228), and the other of only the palatal and approximal portions (Fig. 229, are made, using the moldine in tubes with fusible metal. This will consume only a few minutes. A piece of gold and platinum crown plate, No. 34 U. S. standard gauge, about the length and circumference of the tooth, is then struck up on the palato-approximal cast, from which it receives the palatal and approximal aspects of the tooth (Fig. 230). It is then transferred to the coronal cast, which is previously trimmed the thickness of the gold plate upon the approximal surfaces, and worked down to the exact form of the tooth on the anterior portion. The metal, which is then the exact form of the tooth on all sides, is cut even at the incisive edge, the seam down the front

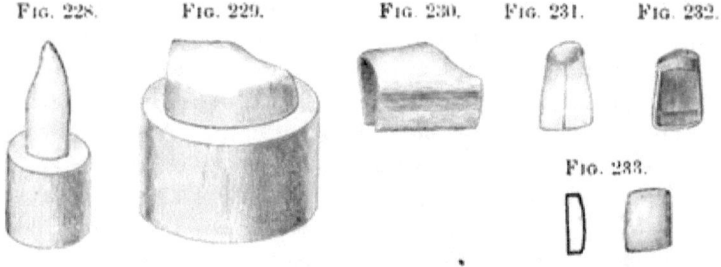

Fig. 228.   Fig. 229.   Fig. 230.   Fig. 231.   Fig. 232.

Fig. 233.

beveled, lapped, and marked, then slightly opened, and the gold slipped off the die. Guided by the mark, the gold is then pressed back to the form assumed on the die, and fitted to the natural crown, which should have been previously trimmed and shaped as described on page 41 (Fig. 20), and the joint on the labial side soldered (Fig. 231). Into the incisive edge, which is open, a narrow strip of gold, about one-sixteenth of an inch in width and thick enough to fill the space, is fitted and soldered. This strengthens the whole crown, and forms a protecting edge for the porcelain front.

We have now a gold contour crown, an exact imitation of the tooth under treatment. The crown is then filled with plaster, and the labial portion ground and filed away, so as to leave the upper part to form the band, and the lower the incisive edge, as represented in Fig. 232.

GOLD CROWNS WITH PORCELAIN FRONTS.

A porcelain tooth which matches in shade and form is then ground and thinned down—in which operation the pins are removed—to the form of a thin porcelain front, thickest at the incisive edge. This front is fitted to replace and represent the labial portion. The porcelain is then removed and a piece of platinum foil adapted to its back, and turned just barely over the edges, the upper and lower of which should be slightly tapered off, as shown in Fig. 233.

The platinum can be made to fit closely to the porcelain by rubbing the edges on a piece of cloth or chamois, on a flat surface. The porcelain front is then adjusted on the crown and cemented on one side with wax, and the crown invested in plaster and marble-dust, so that the seam along the edge of the platinum and gold is exposed on the cemented side and at one end as shown in Fig. 234. When the investment has set, the wax is removed, and prepared gold solder filings, 20-carat fine, are packed in and over the seam, in quantity sufficient to make a

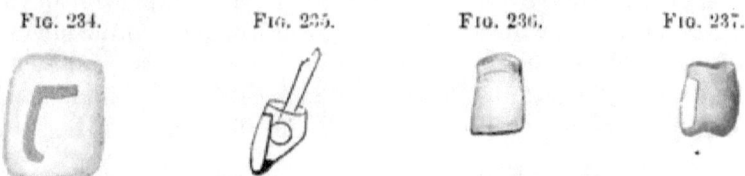

Fig. 234.   Fig. 235.   Fig. 236.   Fig. 237.

perfect joint when finished. The investment is then uniformly heated and the seam soldered. Care must be taken not to flow the solder upon the porcelain, or to use more borax than is absolutely necessary, as otherwise the porcelain will be fractured.

When the investment is cold, the crown is removed, again invested, and the remainder of the seam soldered, or the investment can be immediately removed from the other side, and the soldering completed; though this latter plan is attended with some risk to the porcelain.

After the crown has gone through the finishing process, the excess of porcelain on the inside of the crown is ground away in a few minutes with a small piece of corundum melted on an old oval-shaped bur. The cavity in the crown is, meanwhile, kept filled with water (Fig. 235).

In adjusting the crown, the natural tooth can be smeared with articulating paste, which will easily indicate any point obstructing its perfect adaptation. When fitted, the crown is attached with oxyphosphate cement. Fig. 236 shows the completed crown.

Bicuspids and molars can be made in the same manner by using a contour crown of gold lined with platinum (see chapter on contoured crowns), and, after its adjustment, filling in the crown with plaster, and then cutting away the labial portion and replacing it with porcelain (Fig. 237).

FIG. 238.

Crowning in Cases of Abrasion.—In a case of extensive abrasion of the incisive edges of the anterior teeth, with pulp living though considerably calcified in the coronal section, crown-work to restore the length and form of the teeth is best performed by removing a portion of the labial aspect of the natural crown and then forming the artificial crown similar to a gold collar crown without the pin. An accurately adapted collar, cemented with oxyphosphate, will usually hold the crown securely. Fig. 238 gives an outline of the construction of such a crown. If a case should suggest the necessity of a pin, a short one can be so inserted as not to endanger the pulp.

COLLAR CROWNS HYGIENICALLY CONSIDERED.

The principal argument against ferruled or collared crowns is that they are productive of irritation to the peridental membrane, ultimately causing its absorption and the exposure of the collar. This would be theoretically and practically true of a rough or porous substance encircling the root, or of an imperfectly and unskillfully adjusted ferrule or collar which would by its presence hold a position analogous to a calcareous deposit, but no such comparison can be fairly made with a perfectly fitted collar, forming at its edge a smooth and imperceptible union with the sides of the root, and presenting a uniform and benign surface to the investing membrane. In case of perfectly adapted collars, when any irritation of the membrane exists, it will be found to result from such causes as usually produce it when the natural crowns are present, namely, dental concretions. A tarnished and unclean condition of the surface of the gold of the collar will produce an irritation of the membranes, which is a matter independent of the collar itself, and easily remedied by cleansing and polishing the surface. Where an acid condition of the secretions of the mouth exists, a collar of platinum or iridium, or one of gold and platinum crown-metal, presenting the platinum surface, is suggested in preference to gold, as these metals will not be affected, but will constantly present an untarnished surface.

CHAPTER XII.

SPECIAL FORMS OF GOLD CROWNS WITH PORCELAIN FRONTS.

THE PARR CROWN.

This crown, which in form of construction possesses special points of merit, is largely used by Dr. H. A. Parr.

The root is prepared, banded, and capped without a pin, the same as for a gold collar crown (A, Fig. 239). A hole is made in the cap, and a post fitted in the canal. A piece of gold plate, fully the size of the cap on the root, is adjusted on the post above the cap by making a hole in the gold in which the post will fit tightly. The gold plate is then adapted to the cap on the

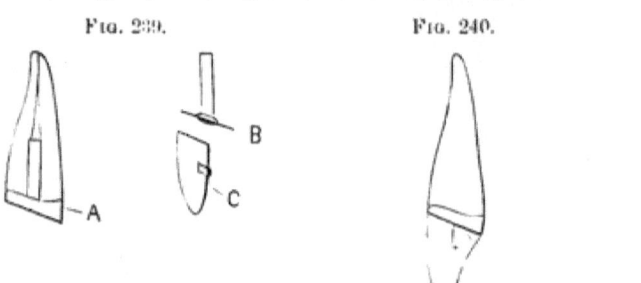

Fig. 239.   Fig. 240.   Fig. 241.

root, and burnished into any open space around the post, forming an outside cap; the pin and the outer cap are then removed and soldered together and adjusted on the inner cap, and the edge of the outer cap trimmed even with it (B). The porcelain tooth to form the crown, C, is fitted and attached to the outer cap, which, when finished, is cemented in position as shown in Fig. 240.

The advantage of this form of crown is that the root is securely and permanently capped independently of the crown, which can be removed without disturbing the cap on the root.

Dr. Parr, in using this style of crown in bridge-work, constructs the outer cap with a band which half encircles the inner cap, and tapers off from the palatal to the labial section, as illustrated in Fig. 241. The cap on the root is cemented with oxyphosphate, and the post and outer cap with gutta-percha.

THE LEECH CROWN.

Dr. Leech's crown is thus described by Dr. J. E. Dexter:[1]

"A method devised by Dr. H. K. Leech, of Philadelphia, shown in Fig. 242, and described in the *Dental Cosmos* for April, 1879, is as follows: The root is drilled out to a depth of about three-eighths of an inch to a diameter of about No. 16, standard (American) wire gauge, the bottom of the hole being flared or enlarged, and the canal above filled with gutta-percha. A gold tube is made to fit the hole accurately and project sufficiently for convenience of handling, and is soldered through a hole in a gold base struck to the root, projecting through the plate some distance. A plate tooth is fitted to the root and plate and soldered to the latter, gold being flowed onto the plate and backing and around the projecting tube to form the palatal contour, and the tube cut off flush with the latter. We now have a plate tooth, gold backed, with a tube-pivot, the orifice of which opens on the palatal aspect of our tooth. The root-end of the tube is now slit perpendicularly in three or four places, for about two-thirds of its length, a thin sheet of warmed gutta-percha is placed on the base of the crown around the tube, and the whole is pushed securely to place. Now pack gold or tin into the tube, *condensing the bottom portions so that the slit end will spread and tightly fill the flared end of the hole in the root*, and the operation is complete."

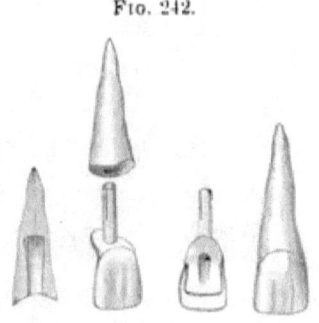

Fig. 242.

A collar crown fastened with a tube-pivot as described can be used to advantage in detachable bridge-work, as the tube if filled with tin foil will admit of the crown being easily detached.

[1] *Dental Cosmos*, May, 1883.

THE LOW CROWN.

In the method for crowning pulpless roots of Dr. J. E. Low, of Chicago, the root-canal is reamed out with an instrument which at the same time shapes the end of the root, or a portion of it, to receive a combined post and cap, which the inventor

Fig. 243.

calls a "step-plug," from its peculiar form. There are seven sizes of the cutting instruments (Fig. 243), and corresponding exactly with them seven sizes of the step-plugs. These step-plugs are not unlike a minute cone-pulley set in a saucer-shaped cap upon the bottom of which is a stout boss. They are made of platinum and nickel. As these last fit the prepared root accurately, it is claimed that they afford a secure foundation for the artificial crown and also prevent longitudinal fracture of the root.

Fig. 244. Fig. 245. Fig. 246.

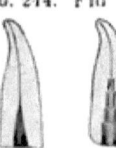  

To describe and illustrate the process, the root of a central incisor (Fig. 244) is selected. The end of the root is first ground level with the palatal margin of the gum. A cutting instrument of suitable size is then selected, with which the root is shaped as shown in Fig. 245. The end of the root is removed enough to permit the palatal edge of the cap of the step-plug (Fig. 246), which is then adjusted, to pass just below the margin of the gum. Fig. 247 shows the step-plug in position, and ready for the adjustment of the porcelain front (Fig. 248) and the construction of the crown, which is completed as in methods previously described. The porcelain front when adapted should meet the labial margin of the gum, the labial surface of the end of the root being trimmed (with the cap in position) with a corundum wheel when necessary to permit it. When the porcelain front is to be adjusted in the mouth, the backing should be warmed, a small quantity of resin and wax cemented upon it, the tooth placed in position in the mouth, and the convex surface of the step-plug cap carefully imbedded in the wax. The wax and the porcelain front should next be carefully removed

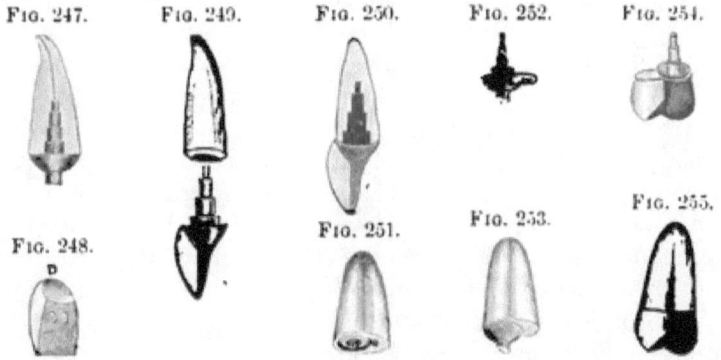

Fig. 247. Fig. 249. Fig. 250. Fig. 252. Fig. 254.

Fig. 248. Fig. 251. Fig. 253. Fig. 255.

and then the step-plug, using pliers for the last. The step-plug is then placed in the wax impression, to which it is fixed with a heated spatula, and invested for soldering. Fig. 249 shows the completed crown ready for final adjustment, and Fig. 250 gives a sectional view of it in position.

In crowning bicuspid roots, one step-plug in the palatal side of the root (Fig. 251) is usually sufficient, the remaining exposed surface of the root-end being covered by adapting thin platinum plate over it and onto the surface of the cap before adjusting the porcelain front (Figs. 252 and 253). Figs. 254 and 255 show the completed bicuspid crown before and after the final adjustment. Molar roots are capped similarly, using two step-plugs.

The plugs are useful in building up badly-decayed roots to support and retain all-gold crowns.

THE PERRY CROWN.

Dr. Safford G. Perry, of New York, employs a porcelain crown in combination with a capped root. A bicuspid will be taken as a typical case to illustrate Dr. Perry's method. The end of the root, by trimming the edge, is given the tapering form shown at A, Fig. 256. The collar (B) is made very narrow. A cap is fitted on the collar, and a post or posts fitted in the roots and through the cap. The entire cap is made of platinum soldered with pure gold. Enough of the post is allowed to extend beyond the cap to attach and firmly retain the porcelain crown (C).

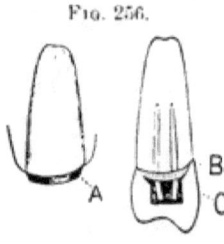

Fig. 256.

One or two holes to serve as vents are drilled through the top of the cap, and it is set in place with oxyphosphate, the excess escaping through the holes. The holes are then reamed out and filled with gold, and the edge of the collar, under the gum, is burnished to the root. The porcelain crown used is similar in principle to the Howland crown, but differs in the details of its formation. The base is given a curve approximating that of the line of the margin of the gum, with the palatal portion projecting slightly above it, to include a little of the cervix. The cavity in the porcelain is given a size, form, and position which will receive the posts extending from the cap without impairing the strength of the crown-walls. Thus they are made round in the incisors and cuspids (A, Fig. 257), oval in the bicuspids (A, Fig. 258), and following the curve of the line of the posts in the molars.[1]

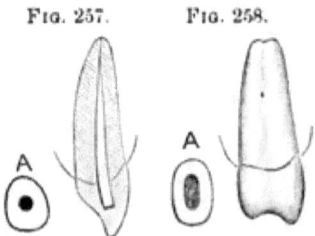

Fig. 257.   Fig. 258.

---

[1] The form of porcelain crown described is also used by Dr. Perry without capping the root, similarly to the Howland crown. In such cases he usually protects the end of the root either with a surface of condensed gold foil anchored in a reamed cavity formed around the post in the root-canal, or with a very thin disk of gutta-percha. The disk with the aid of heat or chloroform is made to form a line of union between the root and crown. In both methods oxyphosphate is used in the cavity which receives the post to attach the crown.

A suitable crown (C, Fig. 256) having been selected, it is ground and fitted in proper position on the cap. This operation is facilitated by perforating a disk of marking-paper with the posts and adjusting it on the cap. Then, as the crown is placed on the cap and pressed against it, points which prevent perfect adjustment are marked on the porcelain. By this means a close joint is easily secured. The edge of the porcelain should be fitted under the free margin of the gum, especially at the cervico-palatal part. The porcelain crown is next set over the projecting pins, and cemented to the cap with oxyphosphate.

The advantage of this method is, that the root being slightly tapered, the collar can be made to fit absolutely, while the excess of oxyphosphate is gotten rid of through the vent-holes, instead of being squeezed out around the edge of the collar. The edge of the collar being made to a knife-edge, can be properly burnished before the crown is placed, so that it will not irritate the gum or make a shelf. The crown covers the cap, and can usually be ground and fitted so as to entirely hide any exposed portion of the collar, the junction of which with the cap should be trimmed and then rounded with a burnisher, to give a form which will better meet the interior of the porcelain cap or crown. Fig. 258 shows the finished crown. Fig. 257 gives a sectional view of a central incisor. The porcelain can be replaced at any time in case of fracture without disturbing the cap on the root. The easy repair thus afforded, the simplicity of construction, and the artistic result, are the special features of this form of crown.

CHAPTER XIII.

CROWNING FRACTURED TEETH AND ROOTS—CROWNING MOLAR ROOTS DECAYED APART AT BIFURCATION—CROWNING IN CASES OF IRREGULARITY.

The crowning of fractured teeth and roots is a process that requires skill and delicate treatment. Its practicability depends on the nature of the fracture, the previous health of the parts, and the length of time that has elapsed since the occurrence of the injury.

LONGITUDINAL FRACTURE OF THE CROWN AND ROOT.

By this is meant a fracture extending lengthwise through the crown or what remains of it, and down the root or roots. Foreign substances having been removed from within and around the parts, the crevice of the fracture is syringed thoroughly with a solution of carbolic acid and tepid water. The fractured parts of the root are then drawn together with waxed floss silk, passed at least twice around the tooth, and tied, the ends being passed through twice in forming the knot. The pulp-chamber is then prepared, and dovetail slots drilled across the parts (Fig. 259). If it is suspected that in the preparation any particles of dentine have invaded the crevice of the fracture, the ligature must be removed, the parts again syringed, and the ligature readjusted. The upper parts of the root-canals are then filled with gutta-percha, and the main body of the cavity and the slots with a hard, quick-setting amalgam. A collar crown should always be used in these cases. If the form to be used has a post, a short, small tube of gold or platinum should be set in the amalgam in proper position to receive it. At the next visit of the patient the ligature is removed and the parts carefully prepared for crowning. The circum-

Fig. 259.

ference of the root is first measured with a wire, a tight-fitting collar constructed, and the crown then completed in the usual manner.

The great drawback in these cases is that the patient generally fails to present himself immediately for treatment, and foreign substances work into the fracture, causing inflammation, which is difficult to control. Often subsequent to treatment a septic condition of the fracture supervenes, the irritation caused thereby and the exudations from the fracture becoming so annoying that extraction is the only alternative.

Teeth fractured as above described are rarely found with living pulps.

FRACTURE OF THE CROWN WITH SLANTING FRACTURE OF THE ROOT.

Fractures of this kind are common, especially in bicuspids, where large fillings are inserted extending from the anterior to the posterior approximal walls, leaving the separated buccal and palatal cusps to bear the brunt of mastication.

In such cases the fracture seldom extends beyond the edge of the alveolar process. The fractured part having been carefully

Fig. 260.   Fig. 261.

removed, a dovetail slot is made in the crown or root, into which gutta-percha is inserted for a day and the membrane of the gum pressed back, so as to fully expose the surface of the fracture. The form of the root is then in a measure restored with amalgam, which, when hard, is polished (Figs. 260 and 261).

The root is then crowned, the mode of operation being the same as in any other case.

CROWNING MOLAR ROOTS DECAYED APART AT THE BIFURCATION.

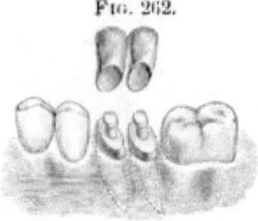

Fig. 262.

The roots of a molar decayed apart at the bifurcation can often be crowned serviceably by making a cap for each root separately, and then soldering the sides of the cap together (Fig. 262). Where one root is missing, the other can be crowned singly.

DR. FARRAR'S CANTILEVER CROWN.

Figs. 263 and 264 represent Dr. J. N. Farrar's cantilever crowns. He describes them as follows: Fig. 263 illustrates a sectional view of three teeth, and an amputated first bicuspid root preserved by a screw, showing the application of the cantilever crown T P, set upon the decayed second bicuspid and

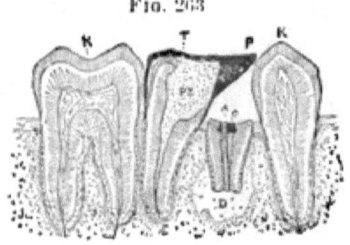
Fig. 263

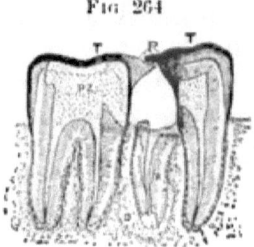
Fig. 264

made to project over to bridge the space formed by the loss of the first bicuspid, and resting in contact with the cuspid so as to connect the broken line of masticating surfaces and prevent tilting forward of the second bicuspid. The abscessed root here shown was extracted.[1] Fig. 264 illustrates the appearance of two molars, the posterior half of one of which is destroyed, showing also the application of two thimble-crowns, which are constructed so as to form a cantilever bridge over the chasm by locking midway in such a manner as to prevent tilting or sliding of surfaces, and at the same time be easily cleansed by a quill or thread.

---

[1] *Dental Cosmos*, vol. xxvi, No. 3.

METHODS OF CROWNING IN CASES OF IRREGULARITY.

Fig. 265 shows a method of treating a case of irregularity without destroying the vitality of the pulp. The tooth at B, which stood inside the line of the lower teeth when the mouth was closed, was trimmed, shaped, and capped. To this cap was attached the tooth at A, with an oval-shaped piece of gold that cleared the lower teeth in occlusion. The cap was then cemented to the natural crown.

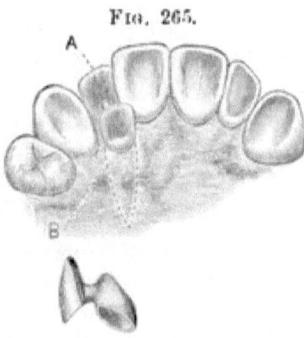

Fig. 265.

Fig. 266 represents a case of irregularity treated by Dr. Bonwill, who says in his description of it,—

"This shows a case of irregularity which was beyond correction, on account of the poor character of the teeth, their position in the palatal arch, and the age of the patient. In such cases I

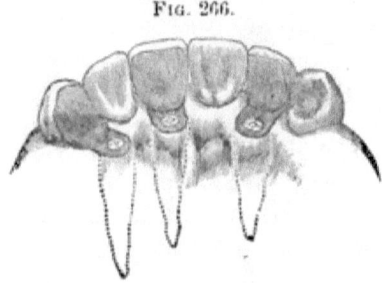

Fig. 266.

do not hesitate to cut off the crown, destroy the pulp, and insert an artificial crown. The crown is brought in the circle and connected with the root by a strip of heavy gold plate. The plate is attached to the root with a post or a screw with a nut."

CHAPTER XIV.

PARTIAL CROWNS.

Gold.—Partial crowns of gold for the protection of plastic fillings in large cavities and for the restoration of contour are often desirable when, for any reason, a solid metallic filling cannot well be inserted. The cavity having been properly excavated, its orifice is trimmed as uniformly straight or circular as its position and character will allow, and the edge of the enamel beveled off, tapering towards the center. In the preparation of cavities in the grinding-surface, trimming and cutting away the enamel should be confined to that surface. In approximal cavities which reach the grinding-surface, it is advisable to extend them in that surface and bring the gold over and anchor it there, so as to afford greater security against its displacement in mastication. Where decay extends close to the margin of the gum, if the tooth is trimmed away so that the gold will extend just under its free edge, a recurrence of decay at that point will be avoided. The bicuspid shown in Fig. 267 will serve as a typical case to illustrate the constructive details.

Fig. 267.

The cavity having been properly prepared, a die of the tooth in its original form is then secured. For this purpose the mold is made by taking an impression of the tooth with wax, making a plaster model, and then restoring the contour and forming from it the mold in gutta-percha or moldine; or the shape of the natural tooth may be restored with wax or gutta-percha and the mold made directly from it in plaster. The die and counter-die having been formed (see article on "Molds and Dies"), a piece of pure gold, No. 28 to 30 standard gauge, the exact thickness being governed by the size and nature of the cavity, is struck up to the form and size of the part to be capped. The gold is

PARTIAL CROWNS.

then adjusted to the cavity, to the margin of which the edges are trimmed and burnished to fit close and flush. In the case of large cavities including a part or the whole of the approximal surface, a model of the tooth and the empty cavity from an impression taken in wax will sometimes facilitate and guide the preliminary trimming and shaping of the gold. Two headed pins fixed on the inside of the cap (Fig. 268) are usually sufficient to secure it, but others can be added if the conditions of the case seem to require it. In compound cavities, including one side and the grinding-surface, one pin at least should be fixed in the latter portion. Where the grinding and both approximal surfaces are included, a wire should be extended from one side to the other (Fig. 269), but the brace should not touch the bottom of the cavity.

Fig. 268.

Fig. 269.

In pulpless teeth the pin from the upper part of the cap should extend up the canal, giving great stability in such cases (Fig. 270).

In soldering the pins or loops when inserted in holes drilled in the gold, a little solder can be flowed over the adjacent parts if deemed necessary to stiffen and strengthen them.

Fig. 270.

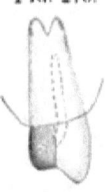

Oxyphosphate is, as a rule, the best to be used in connection with these caps, as it forms a solid and unyielding foundation, and, when properly protected from the fluids of the mouth by a carefully adjusted cap, is very durable.

The cement should be first inserted in the cavity, and then a small quantity placed around the pins of the cap, which should be immediately adjusted accurately in position. When gutta-percha is used, it is heated and applied in the same manner to cavity and cap. The cap is then heated, pressed into position, and held there until the gutta-percha hardens. This can be hastened by the application of cold water from a syringe. The surplus of gutta-percha is then removed, and the edges of the gold burnished.

These caps applied to teeth with living pulps show durability of a commendable character. The advantage they possess over pieces of porcelain is found in the close joint that can be made with the edge of the enamel by burnishing the gold against it.

Dr. H. A. Parr, in this style of work, adapts No. 60 platinum foil to the form of the inner walls of the cavity and just over its edges by the aid of burnishers and cotton twisted on the end of an instrument, assisted by frequent annealing of the platinum. The matrix thus formed is then filled with wax, chilled, and removed from the cavity and invested, after which fine gold or 22-carat solder is melted into it. The plug of gold thus formed is properly trimmed and polished, and cemented in the cavity of the tooth. When completed, it has the appearance of a gold filling. If necessary, the cavity can be previously partly filled with amalgam or shaped with it, to give a better form to permit the removal of the shell of platinum foil. Plugs so made can occasionally be utilized as an anchorage for bridge-work.

Fig. 271.

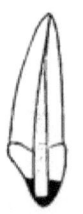

Fig. 272.

Dr. W. B. Ames, of Chicago, makes solid gold tips for abraded pulpless teeth in the following manner for cases in which he prefers not to insert fillings: An opening is made through the occluding surface of the crown into the root-canal. A flat post, wide enough to fit closely in the pulp-chamber across its greatest diameter, thus tending to prevent any rotary motion of the gold tip, is then formed. A very thin piece of pure gold plate, say No. 35 American gauge, is adapted and burnished, with hand-burnishers and Herbst's revolving agate points, into all the irregularities of the abraded surface, and into the orifice of the pulp-chamber. The gold is then trimmed flush and even to the edges, and burnished just over them. An opening is next made in the gold cap, and through it the post is inserted in position. A strip of thin gold plate or platinum foil is next adapted around the tooth, well over and above the edge of the gold cap, and trimmed to the length desired for the tip. With the gold cap and post placed accurately in position, the cavity formed by the strip of plate or foil encircling the tooth is filled with wax cement, and cap, post, and plate or foil removed and invested to the lower edge of the latter (see Fig. 271). The investment is then heated, and 20-carat gold plate or 20- or 22-carat solder melted into the matrix formed by the plate or foil over the cap. The surplus gold is then trimmed to the edge of the abraded surface of the tooth and to the desired form for the tip. When finished, the gold tip is cemented in position with a thin mixture of oxyphosphate. Fig. 272 gives a sectional view of a central incisor tipped in this manner.

Porcelain and Gold.—The partial restoration with porcelain and gold of an incisor crown such as is shown in Fig. 273 is often desirable. The edges of the crown to form the joining with the porcelain are trimmed straight and level, and then polished. A shallow groove is generally formed to advantage at A, Fig. 274. A very thin piece of platinum is then adapted to the crown as shown at B, Fig. 275. The pin C is fitted to the root-canal, passing through the platinum. The post and cap of platinum are then attached with wax, removed, invested, and soldered with pure gold. A little of the gold at the same time is flowed over the cap. The

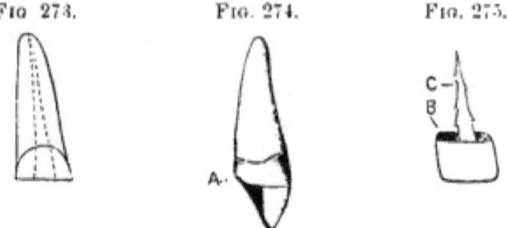

Fig. 273.   Fig. 274.   Fig. 275.

cap and post are then adjusted to the crown, and the cap is trimmed level and burnished closely against the surface of the portion to be restored and into the groove at A, Fig. 274. At this stage of the work, to facilitate the subsequent operations, an impression can be taken which will remove in it the cap, and from this a model can be made. A cross-pin porcelain tooth is then ground down to a size and shape that will properly restore the part and form an accurate joint with the labial edge of the natural crown. The porcelain is then backed, cemented to the cap, removed, and soldered with 20-carat solder. The partial crown when properly finished is cemented in position with oxyphosphate.

In a case such as is represented in Fig. 276 the cap is shaped to the surface of the dentine and enamel at A and over its palatal edge, and the backing on the porcelain is extended out over the palatal edge of the enamel at B. The two sections of the platinum are united in the soldering.

Fig. 277 illustrates a fractured central incisor in which the pulp was not exposed, restored with porcelain by Dr. J. Bond Littig, of New York. The cap to the fractured part was fastened by three small pins as shown in Fig. 278, which illustrates the details of the construction.

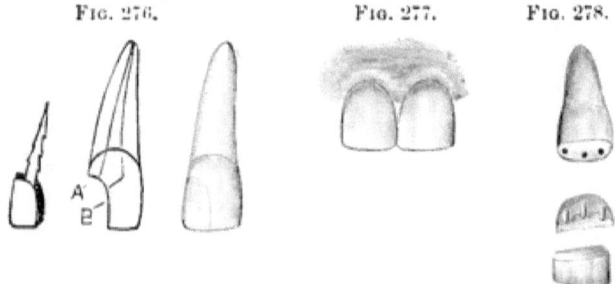

FIG. 276.   FIG. 277.   FIG. 278.

"Where the piece broken off is so narrow that the porcelain tooth cannot be ground to fit in the ordinary way, without cutting out the pins," Dr. Littig describes his method as follows: "First cut a groove in the end of the broken tooth, making slight undercuts. The pins of a suitable porcelain tooth are bent outward, and the ends flattened. The porcelain is then ground away from both ends, until it is made as narrow as the natural tooth is thick or nearly so. The piece is fitted to the end of the tooth by placing the pins in the groove. If the joint is not good, grind away from either tooth or porcelain until it is perfect. Then set the piece with zinc-phosphate, and after it has become hard, grind the tip to shape in the mouth, and polish with moose-hide disk and pumice-stone. Fig. 279 illustrates the

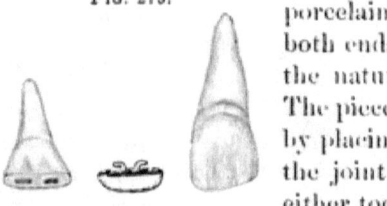

FIG. 279.

second method; the ground porcelain tip, prepared crown, and restored tooth being shown separately." Dr. Littig's third method is to first cap the fractured part with platinum, allowing the ends of the pins, which are soldered with pure gold, to project below the cap. English porcelain body the desired form for the tip is then baked on the cap. The porcelain by this method is secured to the platinum forming the cap, by the platinum pins embodied in it.

Figs. 280, 281, and 282 show how Dr. W. F. Litch's pin-and-plate process may be utilized for the attachment of porcelain tips for broken or decayed incisors, when the appearance of gold fillings is obnoxious to the patient. A represents the por-

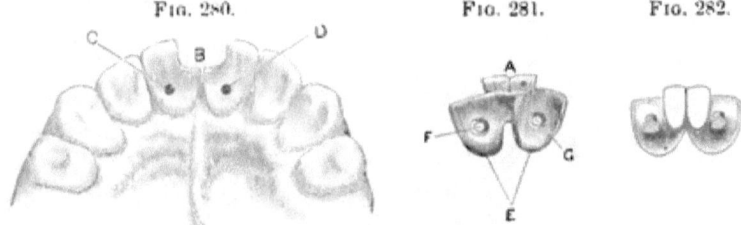

Fig. 280. Fig. 281. Fig. 282.

celain tips; B, the space to be filled by them; C and D, the openings for retaining-pins; F and G, openings in the base-plates (E) for the pins. Fig. 282 shows the appliance with pins attached. Figs. 283 and 284 illustrate a case in which the contour of a single incisor tooth was restored in this manner. Fig. 283 shows the palatal aspect of the tooth, in which the openings for two retaining-pins were drilled, the openings being made quite small. In Fig. 284 is seen the porcelain tip attached to the plate and ready for mounting.

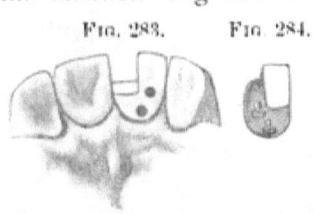

Fig. 283. Fig. 284.

The two retaining-pins will be observed soldered to the plate. In this case the cervical margin of the natural tooth was made level to afford a secure resting-place for the porcelain tip. The appliance has been in use for several months.

Dr. C. H. Land, of Detroit, employs porcelain partial crown-work in cases of the character here described. He first forms a dovetail cavity in the central portion of the section to be tipped or contoured, and then adapts, aided by frequent annealing, a piece of platinum foil (No. 60 U. S. gauge) to the cavity and surface of the part, by means of burnishers, and a pellet of cotton twisted on the end of an instrument. The platinum is then removed, and on its surface porcelain body is placed, and baked in the muffle of a furnace. (Dr. Land's

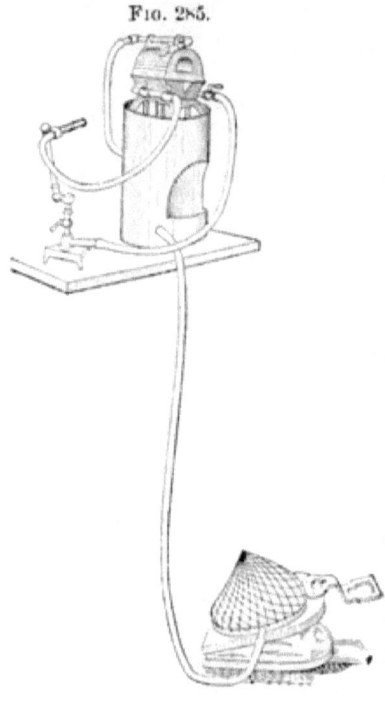

Fig. 285.

Midget Furnace, Fig. 285, is the most suitable for the purpose, as each baking of the porcelain can be performed in about ten minutes.) After the first baking, the thin platinum, which is usually warped by the shrinkage of the porcelain body in the baking, is corrected by readjustment to the tooth, in doing which the porcelain is usually fractured. The interstices and fractures in the porcelain are then filled, the part properly shaped with body, and the porcelain rebaked, and again adjusted and fitted in position. The platinum is next trimmed free of the edge with a corundum-wheel, and the porcelain shaped accurately to the form desired. Any imperfections existing are again filled with body, and the final baking given.

Dr. Land claims that in such cases porcelain, if fitted to the irregularities of the cavity, will be securely retained when cemented with oxyphosphate. An additional attachment is obtained by placing a piece of iridio-platinum wire across the retaining cavity of the tooth, with the ends caught or bent against the sides, and then forming a dovetailed groove across the base of the porcelain tip, which will receive the wire. When a porcelain tip is desired without the platinum base, after the final baking, the platinum is removed by tearing it off the porcelain. When this is the intention, the platinum should not previously be trimmed close, but should be left extending around

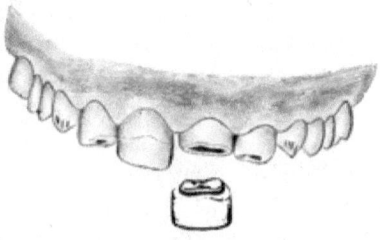

Fig. 286.

and slightly beyond the edge of the porcelain. Fig. 286 illustrates a case of atrophy,[1] in which the tips of the central incisors were contoured with porcelain by Dr. Land. The right central shows the porcelain in position, and the left the porcelain tip ready to be adjusted.

---

[1] This operation was performed by Dr. Land at the meeting of the First District Dental Society of the State of New York, in January, 1889.

CHAPTER XV.

FINISHING AND POLISHING—PROCESS OF CEMENTATION.

FINISHING AND POLISHING CROWN-WORK.

The finishing should be done with small corundum-wheels and points, first coarse, then fine, on the dental engine, which, for this part of the work, is preferable to files or the lathe. In the final finishing use moose-hide points with fine pumice on the engine, or felt wheels on the lathe; and, in the polishing, a brush wheel, with whiting and rouge on the lathe.

The gold should first be properly shaped, which includes trimming the collar off to a fine edge where it fits under the gum, so that the seam of union with the root will be imperceptible. The gold that has been placed on the incisive edge of the incisor and cuspid crowns should be trimmed away, so that although it will protect the porcelain, very little if any will be seen when the crown is in position in the mouth.

INSERTION AND CEMENTATION.

In the insertion and cementation of all crown- and bridge-work, the object to be effected is the same in principle, that is, to form with an insoluble material a solid, substantial, and impervious union between the natural tooth or root presented and the artificial crown. As a cement for this purpose, the plastic oxyphosphate of zinc is generally preferred, and it is in many respects one of the most desirable to be found.

Exact scientific proportions of the chemical substance of which the cement is composed are essential in its preparation, and the oxide of zinc should be in a very fine powder. A standard preparation of the cement should be selected. For all styles of crowns with collars, and for bridge-work depending on crowns of similar construction, slow-setting cement should be used, but for crowns without a ferrule or collar the ordinary cement is preferable. For mixing the cement, a piece of plate-glass about five inches long and three inches wide as a slab, and a small spatula, are suitable. The surface of the slab must be perfectly clean. The acid and powder should first be separately placed on the glass, the amount of powder being fully equal to the requirements of the acid. Should it be found during the mixing that the proportion of powder is too great, the surplus should be instantly thrown off the slab and the mixing, which must be rapid and thorough, continued; the consistence should be that of thick cream. If the slab is placed on a towel which has been saturated with ice-water, the cold will retard the setting, which is quite an advantage in many cases. In this respect the side of a square bottle filled with ice-water and corked is preferable to a slab.

Fig. 287.

The parts to be crowned should be previously syringed with water, then protected by a napkin, bathed with alcohol[1] applied on cotton with tweezers, and wiped with bibulous paper. Each cap or crown, which should have been thoroughly dried, is first filled with enough cement to insure a slight surplus. A small portion is then put in each root-canal or hollow part of a natural crown present, and the artificial crown or bridge immediately adjusted in position. In many cases it is best to quickly remove the napkin and occlude the teeth to insure the occlusion, and then open the mouth and replace the napkin. The crown or bridge should be held under a slight pressure until the cement sets. For this purpose a piece of wood notched on the end or an adjuster (Fig. 287) can be used.

When the cement has set perfectly hard, the surplus around the edges should be removed. In collar or shell crowns the edges of the gold of the collar or shell should be given a final

---

[1] Alcohol used in this manner not only aids in drying the parts, but acts as a styptic on any lacerated portion of the gingival margin.

burnishing. Wet floss silk or dental fiber, charged with pumice, should be passed between and around the teeth to remove every particle of the superfluous cement, and finally the parts should be syringed with tepid water.

The patient should be requested to call in a few days, so that an examination may be made to see if any particles of the cement were overlooked. Cleansing gently at this time facilitates the healing of the gum around the collar or neck. Care in these little details tends to prevent that inflamed appearance and recession of the gum often seen around crowns, and also insures a satisfactory result to the patient and commendation to the dentist.

Previous to insertion the posts of crowns should be slightly barbed. In all-gold cap crowns a vent for the escape of air and surplus cement is usually made in the form of a small hole in the deepest fissure of the grinding-surface. When the cement is hard, the hole must in all cases be closed with a gold or amalgam filling.

When gutta-percha is used for cementing, the cavity in the root and crown having been moistened with chloroform and then dried and heated by a hot-air syringe, a portion of the gutta-percha is inserted and caused to adhere to the sides. The post and the crown are then heated, the proper quantity of the gutta-percha attached, and the crown inserted. When the gutta-percha is cold, the surplus is removed with a sharp instrument, and the edges smoothed by drawing back and forth against them some twisted fibers of cotton saturated with chloroform. Sometimes the post can be fastened with oxyphosphate and the end of the root and artificial crown joined with gutta-percha. In such a case the crown should be heated and the gutta-percha, rolled down very thin, placed on the edges to be united in the form of a perforated disk. The crown is then pressed to position in the oxyphosphate placed in the root. The order of the use of these materials can be reversed where it may be desirable at some future time to easily remove the crown. Bridge-work can be conveniently attached temporarily with gutta-percha. In this event a quantity barely sufficient to fasten the caps should be used. Gutta-percha does not possess sufficient rigidity for general use in bridge-work.

PART III.

BRIDGE-WORK.

BRIDGE-WORK.

The artificial replacement of the loss of a portion of the teeth by bridging the vacant spaces with substitutes, supported in position by means of their attachment to adjoining or intervening natural teeth, is, as we have seen in the introduction, of antique origin, having been practiced long before plates came into use.

Fig. 288.

Fig. 289.

Originally, the application and mechanical construction of such dentures was of a most primitive character; and as the attachments were simply ligatures or clasps of gold, the teeth were more ornamental than useful. Figs. 288, 289, and 290 illustrate the antique methods.[1] Fig. 288 is an illustration of a specimen of ancient Phœnician dentistry. Fig. 289 is that of one in the Etruscan age, dating about five hundred years B.C. Fig. 290 gives a view of the same denture inverted.

Fig. 290.

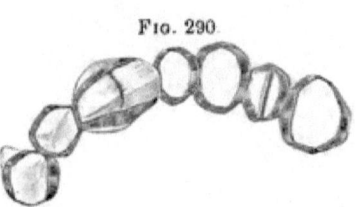

Dentures constructed on the bridging plan by various methods have been occasionally employed from the earliest days of modern

---

[1] See *Independent Practitioner*, vols. vi and vii, "Evidences of Prehistoric Dentistry," by J. G. Van Marter, D.D.S., Rome, Italy. Figs. 288, 289, 290 are copies of the illustrations of the specimens, the first of which is represented as being in the museum of the Louvre, Paris, France, and the second in the Corneto Museum, Corneto, Italy.

dentistry, though until quite recently the system has not obtained general recognition nor been extensively practiced.

Dental literature presents bridging operations as described by J. B. Gariot in 1805, C. F. Delabarre in 1820, Dr. S. S. Fitch in 1829, and Dr. W. H. Dwinelle in 1856. Figs. 291 and 292 are copies of illustrations in Dr. Fitch's work, published in New York in 1829, and Fig. 293 one from a translation of F. Maury's work in 1843. In 1871 the bridging process or bridge principle was again brought to notice by a patent applied for in England by Dr. B. J. Bing, of Paris, for an improved means of supporting and securing a bridge by anchoring with cement or

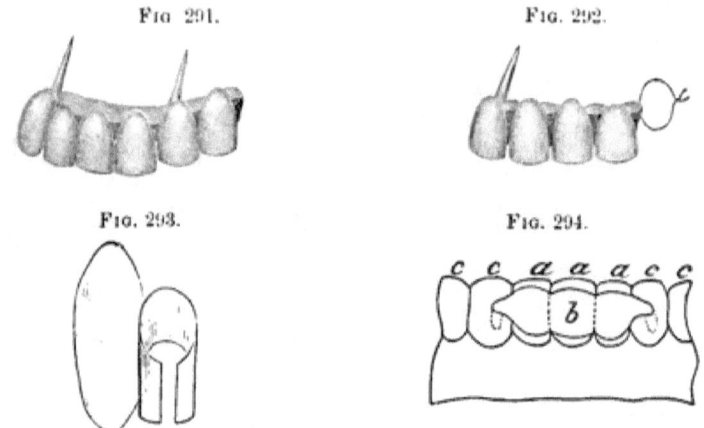

fillings clasps or bars extending from it into holes formed in the adjoining teeth (Fig. 294). The system was also practiced in operations by the late Dr. M. H. Webb, and is described in his "Notes on Operative Dentistry."

The facilities afforded by the artificial crown-work now in vogue for supporting and securing bridge-dentures have caused a revival of the system in an improved form, termed "Bridge-Work," in which artificial crowns cemented to natural teeth or roots are employed as abutments to support artificial teeth which span or bridge the spaces between them. These bridges are so devised, in the best methods, that while supplying the patient with the means of masticating his food the cleanliness of the denture is also provided for.

Ordinarily, bridge-work is immovably cemented in position. The claims set up in its favor are as follows:

First. The perfect replacement of lost teeth by artificial ones, and without the use of a plate.

Second. The absence of any mechanical contrivance to interfere with the tongue in articulation.

Third. The natural teeth are not abraded by the presence of clasps, the functions of the sense of taste are more perfectly performed, and a healthy condition of the tissues preserved, because the gums and palate are not covered over with a plate.

Fourth. The solidity and immovability of the denture at all times, both in speech and mastication.

Fifth. The weight of the denture and the strain of mastication fall on the natural teeth, which are better suited to sustain them than the contiguous alveolar surfaces.

Sixth. Its special adaptation to the replacement of single teeth, or of a small number, where bridge-work is usually superior to any other device.

Seventh. While all operations performed for the restoration of lost teeth, like other remedial operations, are temporary rather than permanent in their results, bridge-work as regards permanency takes equal rank with any other operative procedure.

The following, on the other hand, are the objections raised against bridge-work:

First. It fails to restore the contour of the soft tissues above the bridge, as artificial gums cannot properly be used in this style of work.

Second. The slots beveled under the artificial teeth, called self-cleansing spaces, fill with particles of food.

Third. The speech of the wearer is often affected by these self-cleansing slots under the front teeth.

Fourth. The teeth employed as abutments are usually irreparably destroyed by the process of crowning.

Fifth. If an extensive bridge is made of gold, being immovable, it is impossible to keep it perfectly clean, as the metal will gradually tarnish in parts out of reach of the brush, and will gather offensive matter on its surface and in its interstices.

Sixth. In cases where it becomes necessary to temporarily remove the bridge for the purpose of repair, or because of disease in the teeth which support it, the operation is difficult and the bridge is usually injured so as to unfit it for reinsertion.

Seventh. The teeth which support the bridge are required to bear more force and pressure than nature intended,—where the piece is large many times more,—and, the bridge being permanently attached, at no time can any rest be given the abutments or the contiguous parts by its temporary removal. Thus in a piece of bridge-work consisting of fourteen teeth supported by four natural ones, each one of the natural teeth may have to bear more than three times the strain in supporting the weight of the denture and the force of mastication, that was intended. The ultimate result is evident to any one who is experienced in dental practice; and unless the anatomical conditions are most favorable, the usefulness and durability of such work is decidedly limited in character, considering the time, trouble, and great expense attending it.

Such are the objections which have been put forth against bridge-work; and yet, whatever may be urged against it, its advantages have won from a majority of the profession, including many accepted authorities, an enthusiastic, almost a sensational, indorsement; some practitioners even going so far as to proclaim it the only true method for the insertion of artificial teeth.

Judged impartially, bridge-work has many advantages when practiced by experts who properly construct and apply it. Without doubt it has been abused. Bridges have been inserted where the support was insufficient, or the construction was wrong in principle or faulty from lack of skill. More than this: bridge-work has been passing through the experimental period, when failures are apt to appear more prominently than successes. The chronicles of dental literature, however, in this respect offer only a repetition of the historical difficulties that attend all new departures in the arts.

CHAPTER I.

CONSTRUCTION OF BRIDGE-WORK.

To the skilled mechanical dentist, well versed in metal- and crown-work, bridge-work does not present extreme difficulty. The foundations or abutments—that is, the teeth or roots on which the bridge will rest—are first to be considered, due respect being paid to the mechanical principles controlling the leverage and the force of occlusion in mastication. The amount of strain that can be borne by the different teeth, individually and collectively, according to their position and condition of health, should be carefully calculated. As a rule, the force exerted upon the incisors in occlusion will be directed outward on the upper, and inward on the lower teeth, and its tendency when they support a bridge will be to gradually push them out of line in each direction. When the incisors are replaced by a bridge, the tendency of the force of occlusion is toward a similar result. On the bicuspids and molars the force is direct. The rules which govern the number and position of the teeth or roots that are required as foundations for bridges in practice, are as follows:

One central root will support two centrals, and if spurs or bars from the sides of the bridge rest upon or are anchored in the adjoining teeth, a lateral in addition.

Two central roots will support the four incisors, spurs or bars resting on or anchored in the cuspids to be used additionally, if the case requires them.

The cuspid roots, alone or with the aid of a central root, will support the six anterior teeth.

One molar or bicuspid on one side, and a bicuspid or molar on the other, with one or two roots in an intermediate position, will support a bridge between them.

One right and one left molar, with the assistance of the two cuspids, will support a bridge comprising the entire arch.

A bridge on one side of the mouth can be supported by two

or three teeth or roots on that side. The cuspids always afford the most reliable support.

In general, the application of these principles will cover the subject of foundations, the operator being governed by the exact condition of individual cases. In a bridge of the six anterior teeth on the two cuspids, when the articulation of the antagonizing teeth is close and deep, the strain should be relieved by an additional attachment of the bridge to the teeth posterior to the cuspids.

Fig. 295.

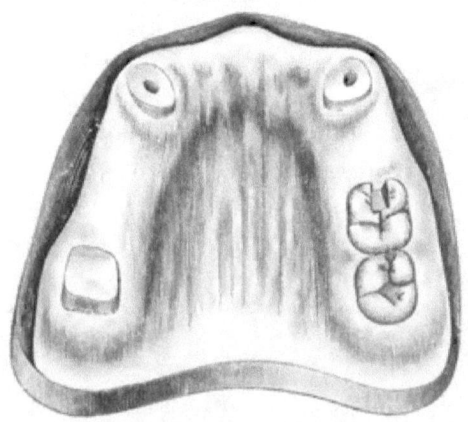

The preparation of teeth or roots to support a bridge is the same as for ordinary crowns, except that the trimming of the sides and the drilling of the root-canals of the various anchorages should be, as far as possible, in parallel lines, so that the collars and posts of the crowns shall move readily to their places in the adjustment of the finished bridge. Teeth or roots which are to be crowned with all-gold cap crowns are crowned by some one of the methods described. Those on which porcelain fronts are to be used are merely capped, the posts being soldered and allowed to project a short distance beyond the caps.

The case represented in Fig. 295 will be used to illustrate the construction of a piece of bridge-work in all its details. The abutments, or supports, consist of the right second molar capped with an all-gold crown, constructed in sections by first forming

the collar and then soldering on the cap (see page 95), the two
cuspid roots capped for collar crowns with porcelain fronts (see
page 89), and the left first molar, which will afford anchorage
to a bar on that side of the bridge (Fig. 296). A slot, dovetail
in form, is usually cut well into the body, but not to an extent
that will endanger the pulp of the last-named crown (Fig. 297).
If the tooth is not decayed, it can be first opened up with a
rubber and corundum disk. The shaping of the slot is best
accomplished with fissure-burs. With the crowns and caps in
position, an impression and articulation of the case are then

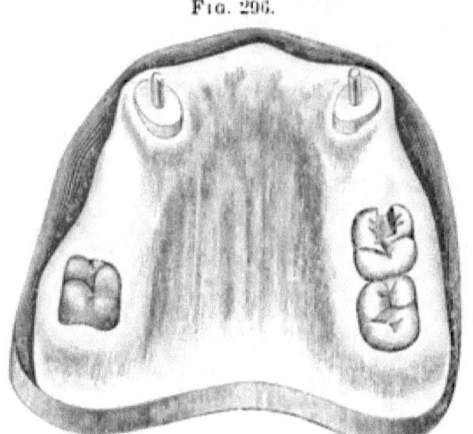

Fig. 296.

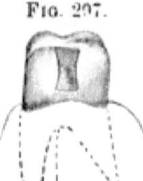

Fig. 297.

taken with plaster slightly colored with carmine. The plaster
is mixed moderately thick and, with the aid of a spoon, placed
around in the mouth on the crowns, caps, and parts to be
included in the bridge, and the antagonizing teeth occluded
tightly and so held until the plaster sets. The mouth is then
opened and the plaster carefully removed, the pieces being
adjusted together should it break. The crowns and caps (the
latter held more firmly by the protruding ends of the pins) are
removed in it. The plaster is then varnished, and, on the side
containing the crowns, a model is run, composed of equal parts
of calcined marble-dust and plaster, to which is added a little
sulphate of potassium,—less than the proportion of salt generally
used,—which causes the mixture to set hard quickly. When the

model has set, it is mounted with plaster on an articulator, and the other side of the colored plaster impression giving the articulation is run with plaster and the opposite section of the articulator adjusted, all at the same time. When the impres-

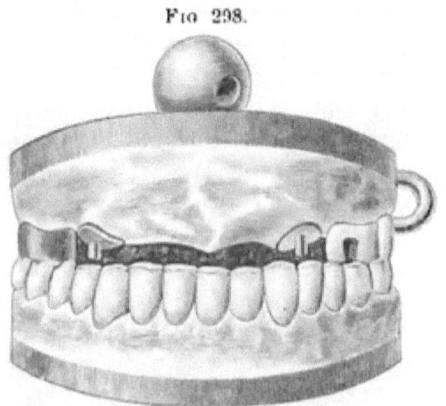

Fig 298.

sion plaster is removed (an operation which is greatly facilitated by its having been colored with carmine), a correct model and articulation of the case will be found, with the crowns and caps in exact position as in the mouth (Fig. 298).

Another method is to first take the impression in an impression-tray, and then the articulation in wax, and make a model and articulation from them in the usual manner.

The pins protruding from the caps on the model are next cut off short. Teeth are selected,—ordinary plate teeth for the incisors and cuspids, and partial teeth, representing the front section of the tooth and styled porcelain facings, which were specially designed for crown- and bridge-work, for the bicuspids and molars (Fig. 299). Cuspids are sometimes used to form the fronts for bicuspids. The teeth are ground and fitted to the model and articulation, so that the labial upper edge of the teeth shall press lightly on the gum. Those which are intended to form the fronts of the caps on the cuspid roots should be adjusted in the ordinary manner for single crowns. To determine the proper positions of the teeth for producing the best appearance, they can be adjusted in the mouth on wax, without the gold crowns or caps of the supports. The correct position of the teeth on the model having been obtained, investing material, composed of one part plaster to two of calcined marble-dust, is placed on the outside of the model on the labial aspect of the teeth, merely sufficient in

FIG. 300.

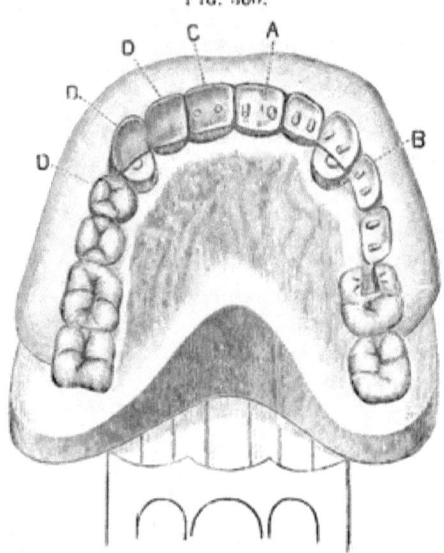

A and B, central incisor and bicuspid ready for metallic backing. C, is a central backed. D, D, D, porcelain fronts as they appear on insertion after the process of backing, capping, and soldering.

quantity to hold them in position, thus forming a matrix. The wax is then removed, exposing the palatal portion, and permitting their form and position to be studied (Fig. 300). The porcelain teeth and fronts, with the exception of fronts for the roots capped, are then removed from their investment, and the base ground from a line on the palatal side below the pins, straight to the labio-cervical edge (A and B, Fig. 300). This is to form the self-cleansing spaces, if they are desired. The incisors are then backed, using either very thin platinum or pure gold (C). The backings are allowed to extend just over the incisive edge as a protection to it, and, if preferred, down on the curve of the self-cleansing space. A more desirable result is secured if the backing extends only to the edge of the self-cleansing space, and the porcelain is polished, as its surface is superior in cleanliness to that of gold.

The bicuspid and molar porcelain fronts, their tips being ground off (A, Fig. 301), are lined in the same manner. A cap of pure gold or gold lined with platinum, representing the grinding-surface of each tooth, is struck up as described and

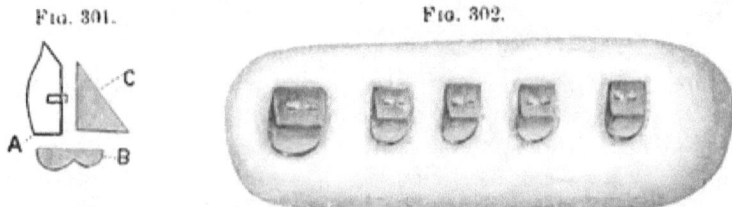

Fig. 301.   Fig. 302.

illustrated in the construction of gold crowns (see page 99), and the concave portion filled by melting in scraps of 20-carat gold plate. The surface is then ground smooth (B), and closely fitted to the tip of the porcelain front to form the occluding surface in accordance with the articulation of the lower teeth, and the space filled in with wax. This metallic occluding surface is to protect the porcelain. Triangular pieces of very thin gold plate or platinum foil (C) are then cut and fitted to the sides, over which they should extend slightly, and the tooth is invested, leaving the back open, presenting the form of a pocket (Fig. 302).

The bar intended to be anchored in the slot cut in the molar on the left side is made of iridio-platinum wire, about No. 15 U. S. standard gauge, with the end shaped as shown in Fig. 303, and fastened with wax to the tooth and cap, and adjusted in the mouth to obtain accuracy of position before soldering.

Fig. 303.

All the teeth should be invested at the same time, the incisors and porcelain fronts to the cuspid crowns as shown in Fig. 304. In the soldering, 20-carat gold plate should be melted into the pockets formed by the cap and side pieces to fill out to the line at A, Fig. 305, and flowed over the backings of the incisors and cuspid fronts in sufficient quantity to shape them as shown at B, Fig. 306. The teeth forming the bridge between the crowns are called "dummies." After the soldering and

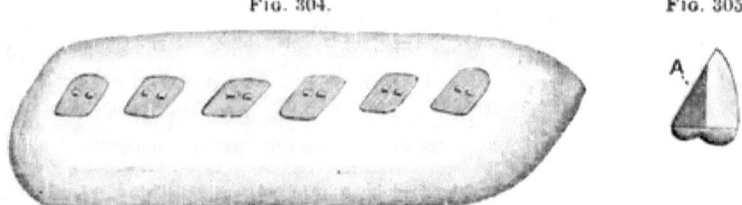

Fig. 304.   Fig. 305.

removal of the borax with acid, each tooth is then, when it is easily done, very carefully trimmed and finished.

The teeth are next placed in their relative positions on the model as shown at D, D, D, Fig. 300, and attached with wax. The model is then detached from the articulator, trimmed down

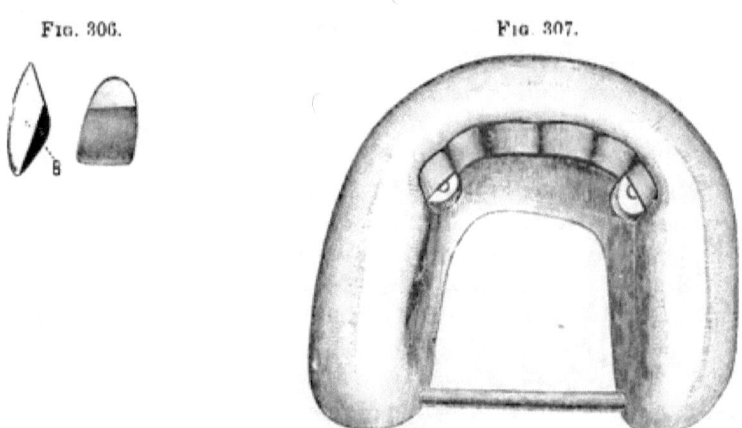

Fig. 306.   Fig. 307.

as much as possible in size, and additional investing material, composed of two parts marble-dust and one of plaster, applied until all of the bridge is covered except the space along the backings and crowns where they are to be united in the soldering. To prevent fracture during the process of soldering, which might readily occur from contraction in so large an investment, an iron wire or a narrow horse-shoe shaped strip of sheet iron should be placed in the investment so as to encircle the teeth and crowns about one-fourth of an inch from their exterior surface (Fig. 307). In the spaces between the backings pieces of gold or platinum plate or wire, about one-eighth of an inch long, are placed lengthwise, and the joints well soldered with 20-carat gold solder. The soldering is best done with a gas blowpipe on a piece of charcoal with a concave depression. When the bridge is removed for finishing, the joints of the backings and crowns are finished with corundum-wheels and points and moosehide points on the engine, and the entire bridge finely polished with whiting carried by a brush-wheel on the lathe. Any little pits that may exist can be filled in with gold foil. The bridge is then ready for insertion (Fig. 308). If the constructive details have been properly performed as described, a finished piece of bridge-work is the result.

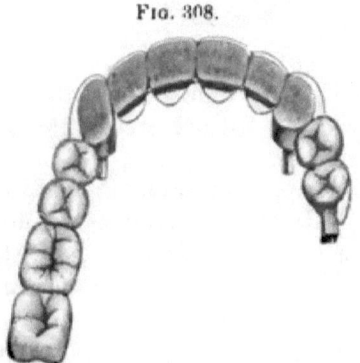

Fig. 308.

In constructing bridge-work many prefer, after the porcelain fronts are backed and the caps forming the occluding surfaces of the bicuspids and molars are properly adjusted on the model, to invest and do the entire soldering at once. When this plan is followed, pieces of gold wire should be laid lengthwise in the slots under the gold caps of the porcelain fronts, and the parts filled in and all the sections of the bridge joined together in the soldering. By this method there is less liability of fracturing the porcelain fronts, but the finishing of the bridge is not so easily or so perfectly done.

In large pieces of work there is some liability to warping, which may be avoided, whichever method of soldering is adopted, by first removing, in proper position, the "dummies" and soldering those of each span together. The spans are then replaced in the matrix and soldered to the abutments.

Adjustment and Attachment.—The bridge when finished is adjusted in the mouth, every point carefully examined, and any alterations required are then made. Should the edges of the collars of any of the crowns catch, so as to prevent their being

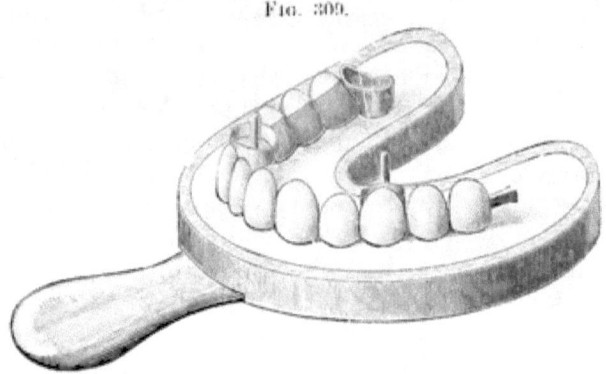
Fig. 309.

placed in position, a small quantity of articulating paste should be applied and the point found and trimmed off. If extensive warping has occurred in the soldering, the bridge must be sawed apart in one or two places, adjusted in the mouth, and removed in an impression-tray, using only sufficient investing material (equal parts of plaster and marble-dust with the usual quantity of sulphate of potassium) to cover the points of the teeth and crowns (Fig. 309). The inner surface of the tray should be oiled.

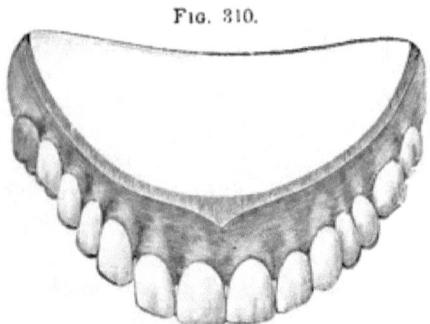
Fig. 310.

The bridge and the investing material are then removed together from the impression-tray, and more investing material is added to complete the investment. The cut parts of the bridge are then soldered together.

When the adjustment of the bridge is accomplished, it can at first be temporarily attached with gutta-percha if desired. For its permanent attachment the pins or posts of the crowns are barbed, and the teeth and roots to which crowns have been fitted are then treated the same as single crowns, and the bridge cemented on with a slow-setting oxyphosphate cement (see page 141). The end of the bar is anchored in the slot by either a gold or an amalgam filling. Fig. 310 represents the bridge in position.

The Construction of Small Pieces of Bridge-work is much simplified by the following method: Crowns are first made for the teeth or roots that form the abutments and temporarily placed in position.

Fig. 311.

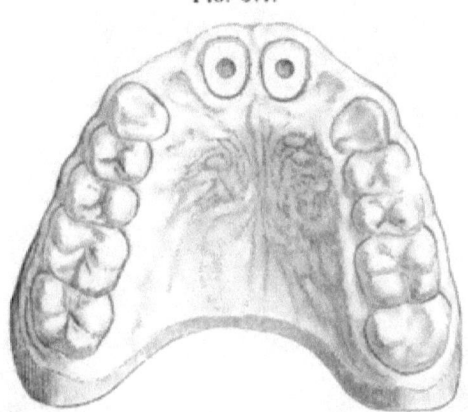

Fig. 312.

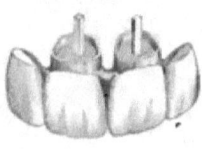

Fig. 313.

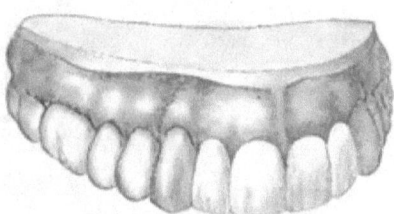

The teeth—"dummies"—which form the span having been ground and backed, are adjusted and cemented with resin and wax in proper position between the crowns. The crowns and dummies are then removed together, in an impression-tray filled with investing material. The inside of the tray should be previously coated with a film of wax with a serrated surface. The impression-tray is then heated, and the investment with the crowns

162   ARTIFICIAL CROWN- AND BRIDGE-WORK.

and dummies removed; more investing material is then mixed, and the exposed parts of the crowns and teeth covered. The investment, when set, is then cut away sufficiently to expose the

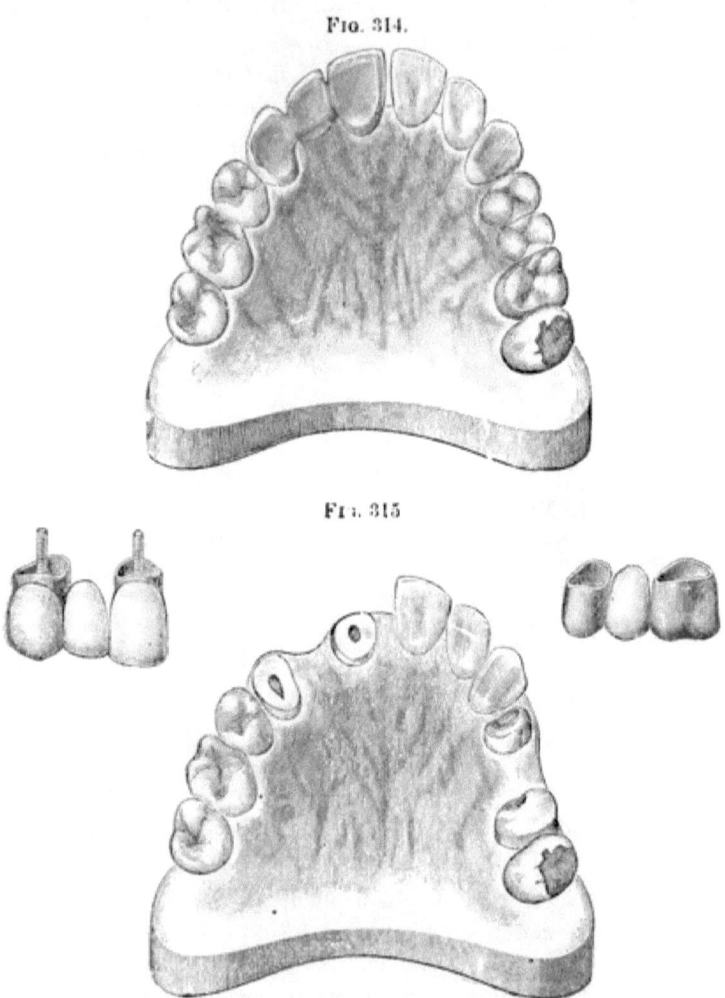

Fig. 314.

Fig. 315.

parts that are to be filled in and united together in the process of soldering. Figs. 311, 312, 313, 314, 315 represent cases of bridge-work constructed in this manner.

CHAPTER II.

SPECIAL PROCESSES AND APPLIANCES IN BRIDGE-WORK

Shoulders on the Anterior Teeth are sometimes desirable, especially on the superior cuspids at the point of occlusion with the lower teeth. A shoulder can be made by melting gold plate into the form of a small ball or globule, then flattening it out and soldering it against the backing.

Another method is to attach with wax transversely across the backing in proper position a strip of gold plate as shown in Fig. 316, and then flow in gold to the line A, by specially investing or in the soldering of the bridge.

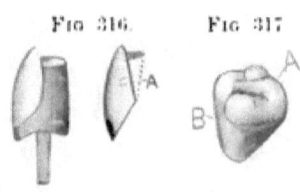

Solid Gold Crowns.—In a close articulation, when the patient prefers strength to appearance, the bicuspids and molars should be made of solid gold. They are constructed by stamping up the cap (A, Fig. 317) representing the grinding-surface (see page 97), and then adjusting and cementing with wax a piece of plate cut and shaped to form the front and sides (B), which is then invested and filled in with gold solder, or, if pure gold has been used in the sections of the crown, with 18- or 20-carat plate.

Seamless contour crowns can be used for the purpose as follows: The proper crowns having been selected, the gold is trimmed and the crowns adjusted in position on the model. A matrix of plaster is then placed on the labial side, which permits the palatal portion to be studied, the crown removed, and cut away to the form required. They are then removed, invested, and filled in with gold plate.

A bar-bridge can be made with these by passing an iridio-platinum wire through the crowns before filling in (Fig. 318).

A Solid Gold Crown for a Pulpless Molar, supporting the end of a bar as shown in Fig. 319, is constructed as follows: The natural crown is ground down, banded, capped, and pivoted as in Fig. 320. The gold forming the top of the cap is made perfectly flat and left projecting a little at the sides. A contoured

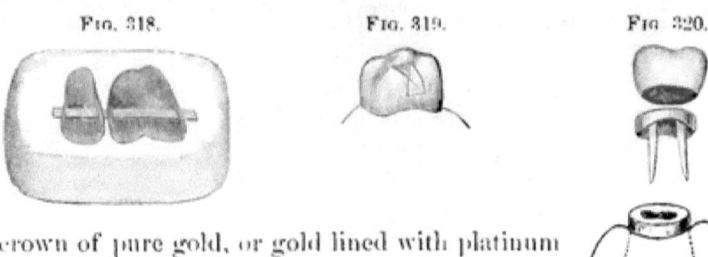

FIG. 318. FIG. 319. FIG. 320.

crown of pure gold, or gold lined with platinum as a precaution against melting, is shortened sufficiently to represent the absent coronal section of the tooth, and with a corundum-disk the orifice of the anchorage cavity is formed. A shell of thin platinum of the size and shape of the anchorage cavity is then inserted and cemented with wax on the inside of the crown. The crown is then invested and filled with 18-carat gold solder, which should be cut into small pieces, and be placed successively in the crown

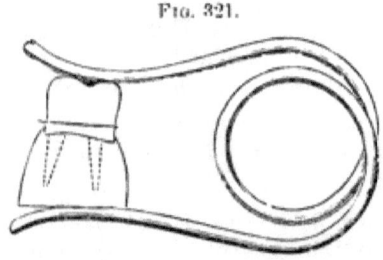

FIG. 321.

and melted by maintaining a uniform heat with the large flame of a blow-pipe. The base of the crown is next ground level and accurately fitted to the cap and articulated to the antagonizing teeth. It is then clamped in position to the cap, the pivots of which are protected with investing material (Fig. 321), and soldered, making a perfect joint. This is an easy method of constructing an otherwise difficult form of crown.

SPECIAL PROCESSES AND APPLIANCES IN BRIDGE-WORK. 165

Fig. 322 illustrates a bridge supported by a bar-anchorage in a solid gold crown on the roots of a molar and a shell crown on a cuspid.

Connecting Bands or Bars for Bridges, which obviate the removal of crowns of intervening natural teeth between the sections of a projected bridge, are formed by passing a heavy band of oval-

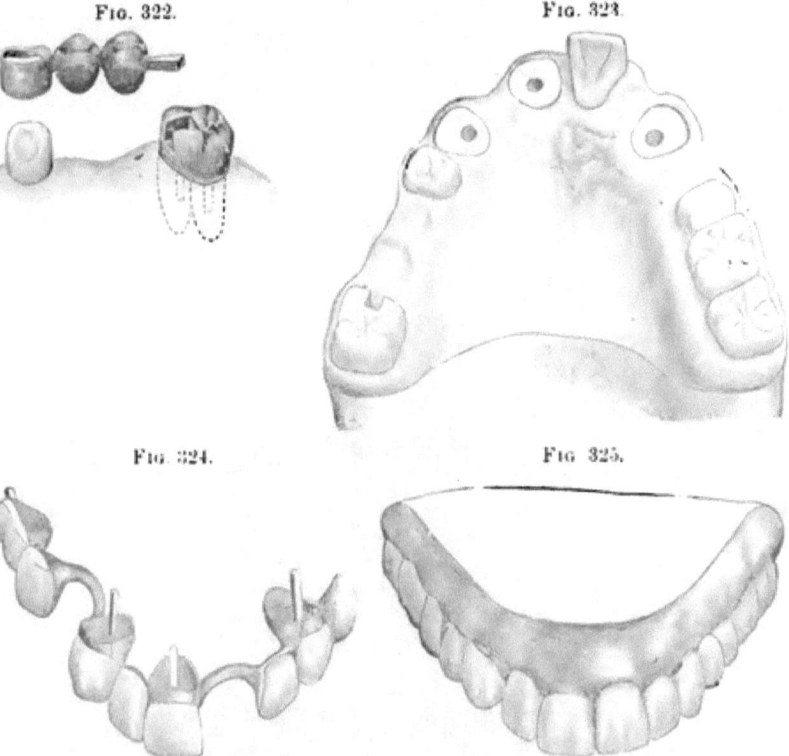

Fig. 322.   Fig. 323.

Fig. 324.   Fig. 325.

shaped gold or iridio-platinum wire around the intervening teeth, close to but not touching them, and resting lightly on the gums. The application of this device was first illustrated by Dr. J. L. Williams, Figs. 323, 324, and 325.

Roots intervening between the abutments of a projected bridge should not be extracted, but whenever practicable they should be treated, filled, and trimmed level with the gum, as they can

166   ARTIFICIAL CROWN- AND BRIDGE-WORK.

usually be made to afford some support for the bridge, which may rest upon them. Figs. 326 and 327 represent a case in which the root of the cuspid on the right side has been so treated and utilized.

Fig. 326.

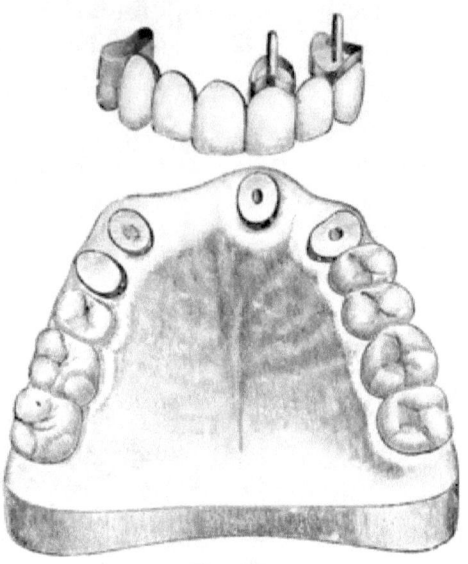

Fig. 327.

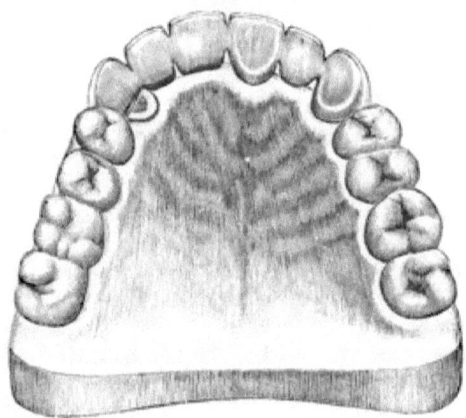

Shell Anchorage or Crown.—The construction of a shell anchorage for a cuspid and its practical application in supporting a bridge cannot be better described than in the following words of Dr. J. L. Williams:

"Fig. 328 shows a piece of work made for a case of quite frequent occurrence. It represents the restoration of the inferior bicuspids and first molar of the right side. A gold crown is made for the second molar, and the three intervening teeth or 'dummies' are then constructed. For the support of the anterior end of the bridge, the method hitherto practiced has been

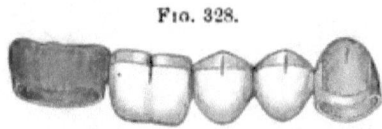

FIG. 328.

to excise the crown of the cuspid and fit a porcelain crown with gold backing to the root, and to this the anterior end of the bridge is soldered.

"Fig. 329 illustrates a device which obviates the necessity for removing the cuspid crown. A gold band is fitted around the cuspid at the front, shown at *a*. This band is allowed to pass a little beneath the margin of the gum, so as to make the smallest possible exhibition of gold. On the lingual aspect

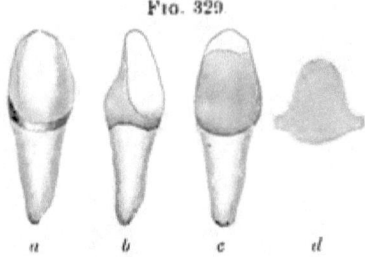

FIG. 329.

a   *b*   *c*   *d*

of the tooth the band is allowed to be nearly the length of the crown. It will be seen that when this band is fitted as perfectly as possible there must necessarily be quite a vacancy between the upper part of the lingual surface of the tooth and the band. It is important that this portion of the band fits the tooth perfectly, and an accurate adaptation is obtained as follows: A piece of very thin platinum or pure gold, rolled to No. 35 American gauge, is fitted over that portion of the lingual surface of the tooth which it is to cover. *d*, Fig. 329, shows the shape that this little plate usually assumes. It can be perfectly fitted by the use of a burnisher, and then, with the band in position, a drop of melted resin wax is flowed into the vacant space between the pure gold and the band. It is now removed from the tooth, invested, and after melting out the wax solder is flowed into the

FIG. 330.   FIG. 331.

vacancy, filling completely the space occupied by the wax. The top of the lingual portion will now be thicker than is necessary, but it can be ground or filed down to the proper thickness. We now have a band which fits all portions of the tooth perfectly. The anterior end of the bridge is soldered to this band, and after the work is properly finished it is cemented in place in the usual

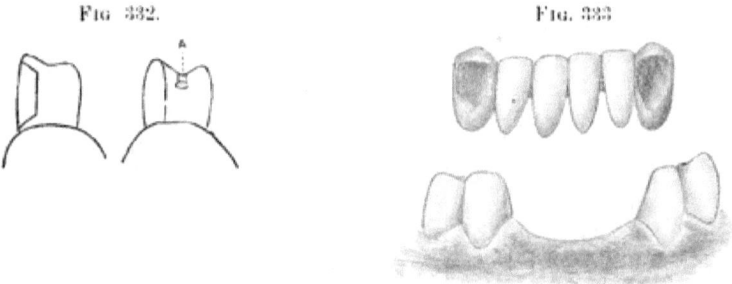

FIG. 332.   FIG. 333.

manner. *b* and *c*, Fig. 329, show side and lingual views of this band after the fitting is completed."

A *Seamless Shell Anchorage* is formed as follows: From an impression of a natural crown taken in plaster, gutta-percha, or moldine, a die in fusible metal is formed, and from it a counter-die in lead is made. (See "Gold Seamless Cap Crown.") On the die a crown is stamped from a seamless cap of gold (Fig. 330). This crown is then fitted on the natural tooth, the labial aspect of which is exposed by the removal of the section of gold covering it (Figs. 331 and 332). A shell for a cuspid can be made from a gold collar as well as a cap. The shell formed in either manner is then filled with investing material, and strengthened by flowing 20-carat solder over the surface.

This process for cuspids has decided advantages for the easy formation of a perfect-fitting crown or shell for bridge-work.

Fig. 332 shows the forms usually given shell crowns for bicuspids. When the gold is removed at the labio-cervical part, the crown should be additionally secured by a pin introduced and soldered at the point A.

Fig. 333 illustrates a case in which the lower incisors are supported by shell crowns on the cuspids.

CHAPTER III.

EXTENSION BRIDGES.

This term is applied to bridges which are chiefly supported by one abutment. In relation to the anterior teeth, it consists in attaching a dummy to an artificial crown, to replace an adjoining absent tooth. A bridge of this style replacing two or three of the posterior teeth is formed by using two of the teeth

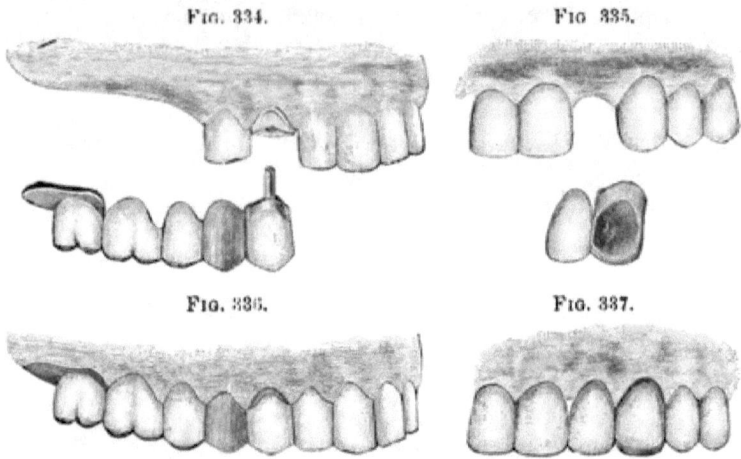

Fig. 334.   Fig. 335.

Fig. 336.   Fig. 337.

anterior to it as one abutment, with a saddle for the other. A saddle is an oval-shaped piece of gold, of the form of the gum and a little larger than the base of the tooth, placed under the posterior tooth of the bridge.

Figs. 334 and 335 represent an extension bridge. A crown on the cuspid, an all-gold crown on the bicuspid, and a saddle, are the abutments. In constructing this bridge, the teeth forming the abutments were first crowned. The crowns were then adjusted in position, and an impression and articulation taken in plaster, in which the crowns were removed. From this impression a model was made of plaster and marble-dust, and an articulation in plain plaster. With the crowns in position on the

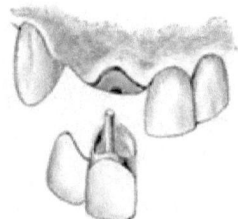
Fig. 338.

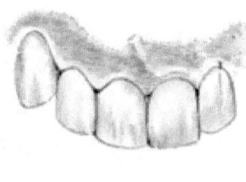
Fig. 339.

model as in the mouth, the bridge was then constructed by the methods described on page 156. The part of the model on

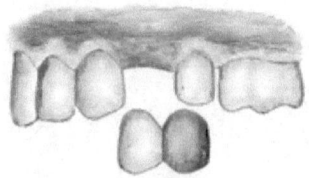
Fig. 340.

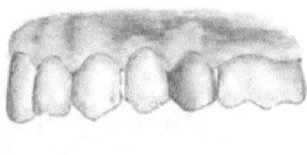
Fig. 341.

which the saddle rested was marked, and enough of the surface of the plaster removed to cause the saddle to press tightly

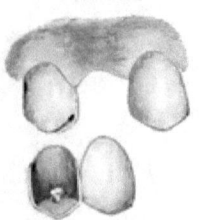
Fig. 342.

against the soft tissues when the bridge should be completed and inserted in the mouth.

A shell crown on a cuspid can be used as an abutment in this

172   ARTIFICIAL CROWN- AND BRIDGE-WORK.

style of bridge, instead of excising the natural crown and mounting an artificial crown on the root for the purpose (Figs. 336, 337).

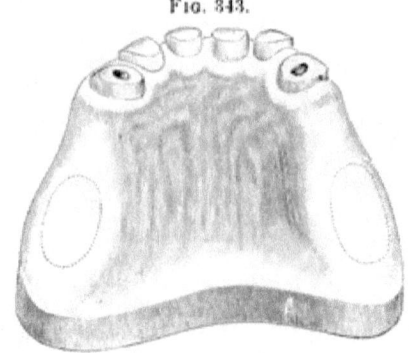

Fig. 343.

Figs. 338, 339, 340, and 341 represent small extension bridges of frequent construction, and Fig. 342 a pin-shell crown which

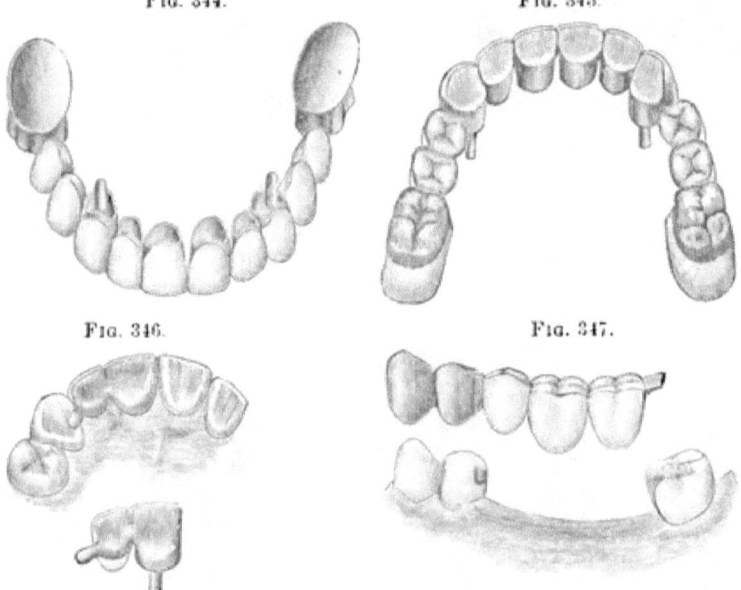

Fig. 344.   Fig. 345.

Fig. 346.   Fig. 347.

can be made to support a porcelain front representing a cuspid or bicuspid.

Figs. 343, 344, and 345 represent an extension bridge constructed by Dr. H. A. Parr, of New York. The anterior abutment consists of the six front teeth, which were all crowned and joined together, the pulps being preserved in the incisors. A saddle on each side forms the posterior abutments.

A Spur Support consists of a flange formed at the end of an extension bridge, affording support by resting on the palatal wall of either an incisor or cuspid, or in the sulcus between the cusps of a bicuspid or molar. Fig. 346 illustrates a case with a

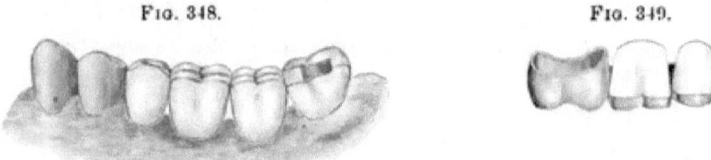

Fig. 348.   Fig. 349.

spur resting against a cuspid. In the case represented in Figs. 347 and 348 the spur rests in an indentation in an amalgam filling in the molar, and is termed a cantilever bridge. A bridge formed with a spur which is anchored in a filling is practically a one-bar bridge (Fig. 349).

CHAPTER IV.

DOUBLE BAR-BRIDGES.

In this style of bridge the teeth or dummies forming it are supported by bars anchored by fillings in the natural teeth forming the abutments. Its use is confined to the insertion of one or two teeth. In the incisors and cuspids the cavities of anchorage are formed in the palato-approximal portion of the teeth contiguous to the space to be bridged, about one-third of an inch from the gingival margin (Fig. 350), and having direct access into at least one of the cavities through the palatal wall.

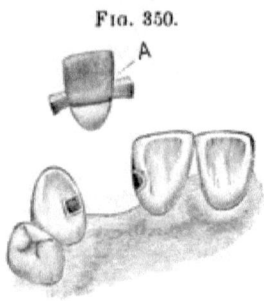

Fig. 350.

In constructing the bridge the bar is first made. Iridio-platinum wire about No. 14 to No. 16 U. S. standard gauge is preferably used for the purpose. The extremities are formed square or triangular, increasing in size towards each end, and fitted deep into the cavities of the teeth which are to support them. The bar is then adjusted and a suitable plate tooth ground and fitted in proper position against it,—a portion of the labial surface of the bar being removed to receive it. The tooth is then backed with very thin platinum, cemented with wax to the bar, and the tooth and bar removed, invested, and soldered. Sufficient gold should be added to properly contour the part (A, Fig. 350). Gold and amalgam are the only filling-materials suited for anchorages of this character. Amalgam is objectionable only when the position of the cavity renders it visible. It can, however, when set be partially removed at the exposed portion and covered with gold. Fastening one end of the bar temporarily with oxyphosphate while the other is being secured, will sometimes facilitate the operation of anchoring. Frequently in the bicuspids and molars it is advantageous to first insert one or both of the fillings, and then drill out sufficient of it to admit the bar, which can then be secured with additional filling-material. When gold is the filling-material used, the rubber-dam is first adjusted on the natural teeth and the bridge placed in position over it.

The practical success of this style of bridge-work depends chiefly upon the character of its supports and the skillfulness with which it is anchored.

A Bar-Bridge with a Detachable Porcelain Front affords access to the cavities of anchorage and admits of replacement of the porcelain in case of fracture.

A modification of Dr. I. F. Wardwell's method, which is simple in construction and application in comparison with most forms in use, is as follows: A thick, narrow piece of 18-carat

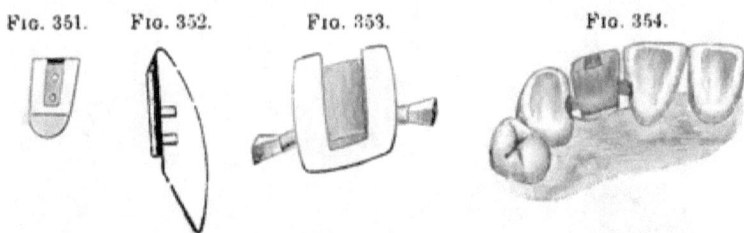

FIG. 351.   FIG. 352.   FIG. 353.   FIG. 354.

gold plate, at least No. 18 standard gauge, is soldered to the tooth and its two sides undercut with a very thin separating file (Figs. 351, 352). A very thin piece of platinum, covering the whole back of the tooth, is then burnished against it, well into the undercuts, the platinum being annealed several times during the operation. The platinum is next held in a flame while a small quantity of pure gold is flowed over the outer surface and then refitted to the back of the tooth, to which it is again burnished. This operation is repeated until the platinum and gold form a moderately light backing which fits perfectly. The platinum surface is then covered with investing material, and on the other side 18- or 20-carat gold plate is flowed until a suitable thickness is obtained. When trimmed into proper shape and attached to the bar, this forms a substantial backing or socket (Fig. 353) in which, when the bar is anchored, the porcelain tooth can be fastened with a little oxyphosphate cement or gutta-percha (Fig. 354).

The Low porcelain front consists of a porcelain facing which fits into a metallic socket, where it is retained by grooves on the sides. The metallic part is soldered in position and the porcelain front cemented on. (See page 126.)

Fig. 355 represents a double bar-bridge formed with an all-gold molar crown.

Dr. J. G. Morey's method of constructing a double bar-bridge with a removable molar or dummy is as follows:

The bar is formed as seen in Fig. 356. A countersunk molar is ground and shaped as seen at A and B, Fig. 357. To the base of the molar and up in the slot is shaped and fitted the shell C,

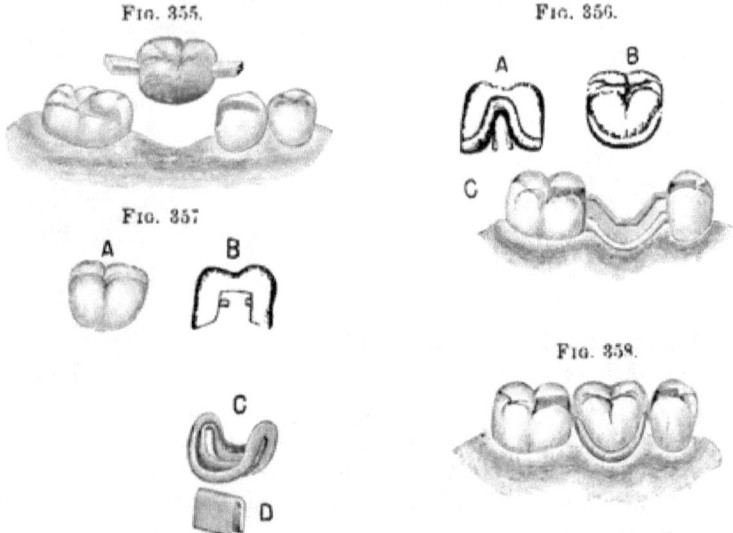

made of a very thin piece of gold and platinum crown-metal by first stamping it on a fusible metal die of the base and then adapting it to the part. A thin piece of clasp-metal (D) is then formed to fit over the bar and in proper position into the slot of the shell (C, Fig. 356), to which it is soldered by investing and soldering from the inside of the shell. The shell is then cemented with oxyphosphate onto the base of the molar as seen at A and B, Fig. 356, and is secured to the bar by springing together the edges of the metal at C. Fig. 358 illustrates the bridge in position.

CHAPTER V.

EXTENSIVE APPLICATIONS OF CROWN- AND BRIDGE-WORK.

The following illustrations of bridge-work by Dr. H. A. Parr, of New York, show to what an extent the system can be applied.

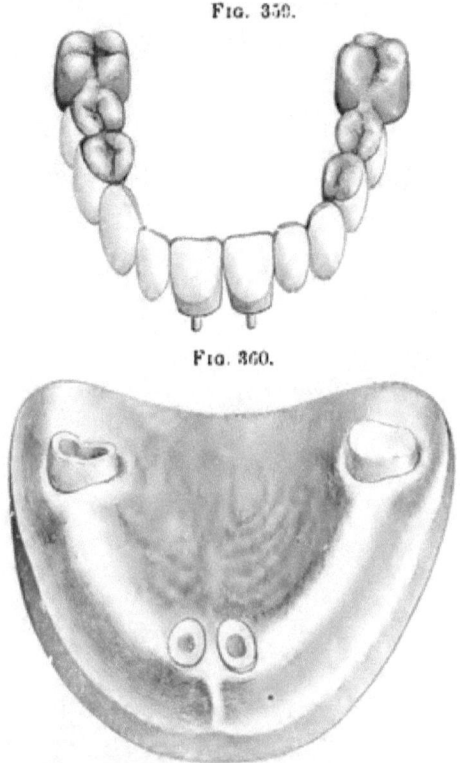

Fig. 359.

Fig. 360.

Figs. 359, 360, 361, and 362 represent a case in which the roots of the two superior centrals, a partially decayed right first molar, and a badly decayed, pulpless left first molar were all that

178 ARTIFICIAL CROWN- AND BRIDGE-WORK.

remained of the upper natural teeth. On the two central roots were mounted collar crowns, and on the two molars all-gold

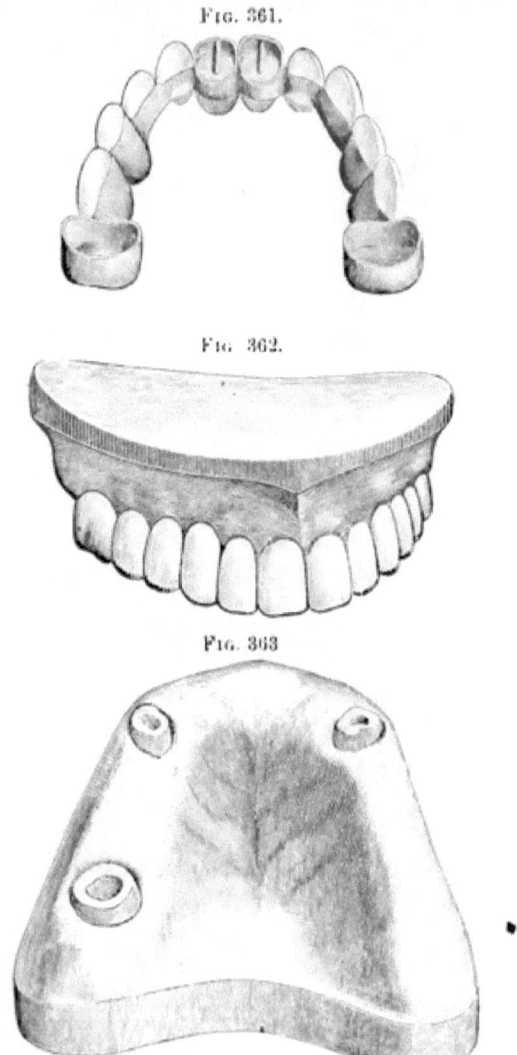

Fig. 361.

Fig. 362.

Fig. 363.

cap crowns. These four crowns, acting as abutments for the bridge denture, bore between them, proportionately on each side, the force and leverage of occlusion. The contour of the arch in the region of the cuspids was restored by a skillful

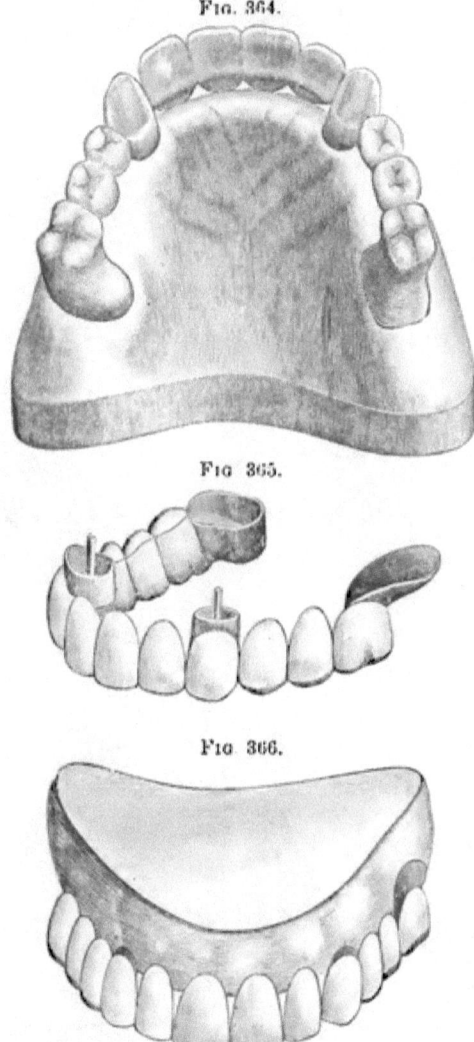

Fig. 364.

Fig. 365.

Fig. 366.

and artistic placing of the artificial teeth, which are prominent and long.

180   ARTIFICIAL CROWN- AND BRIDGE-WORK.

Fig. 367.

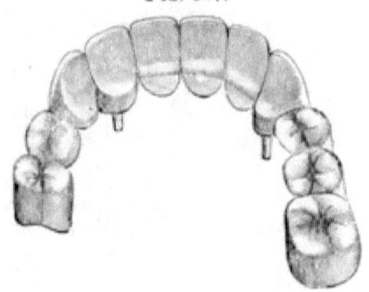

Fig. 368.

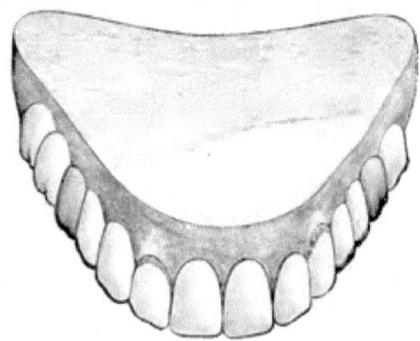

Fig. 369.

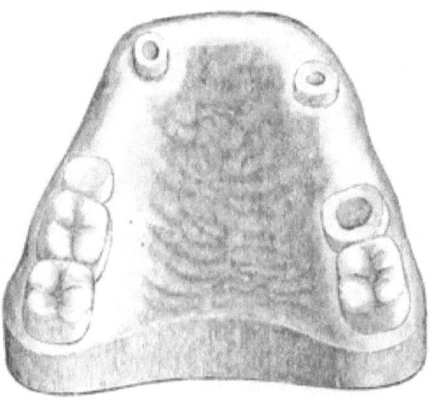

Figs. 363, 364, 365, and 366 represent a case in which two large and firm superior cuspid roots, and a right pulpless molar, with a saddle—an invention of Dr. Parr's—on the left side support a large bridge.

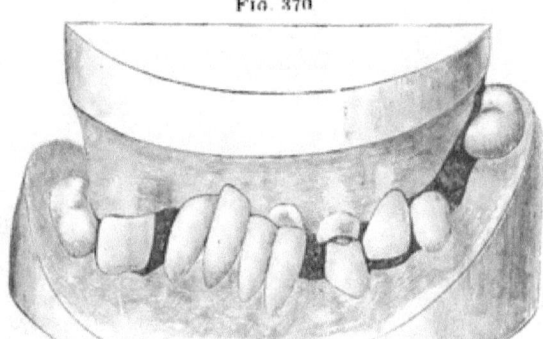

Fig. 370

In the bridge-work illustrated in Figs. 367, 368, and 369, crowns on a second bicuspid, a pulpless molar, and the roots of a cuspid and lateral constitute the abutments.

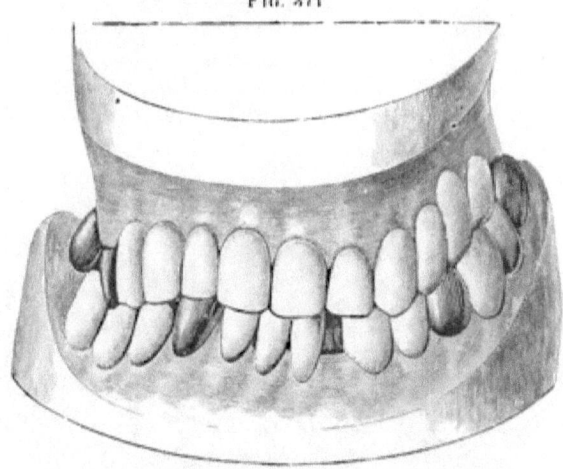

Fig. 371

Figs. 370 and 371 illustrate an extensive case of artificial replacement by crowning and bridging operations. Fig. 370 represents the case as presented for treatment. The few

Fig. 372.

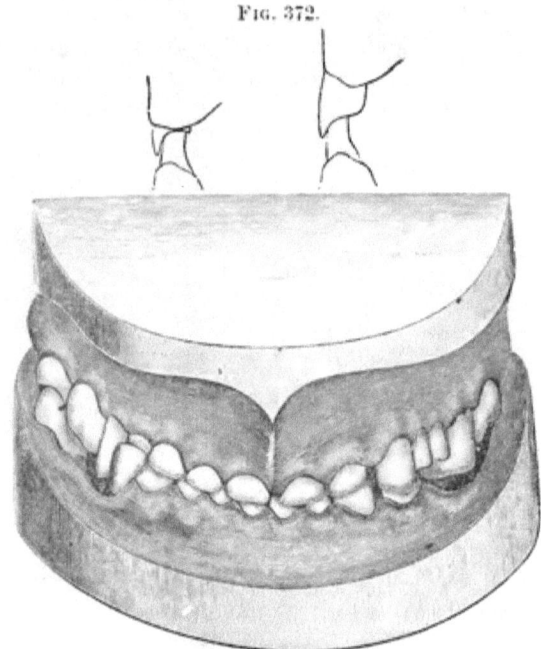

Fig. 373.

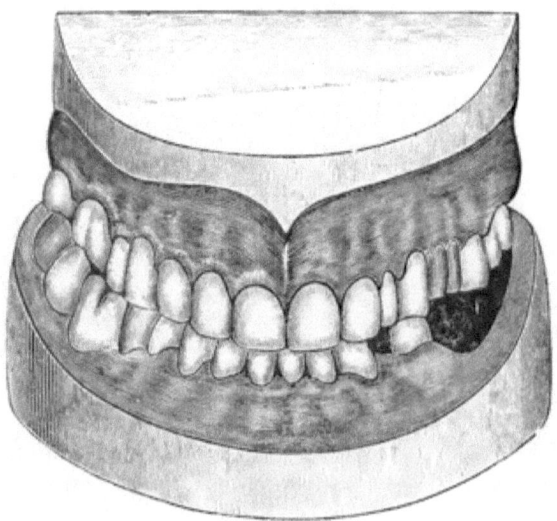

remaining superior and inferior natural teeth had no corresponding antagonists, which caused the interlocking and abnormal condition in regard to occlusion shown. The superior right bicuspid, the left central, and the left cuspid were crowned,

Fig. 374.

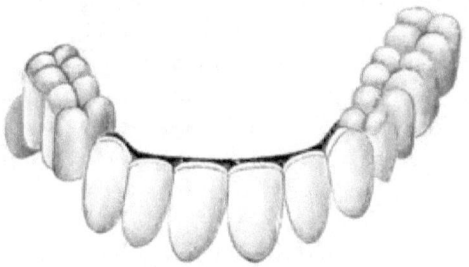

the bicuspid being shortened and the cuspid lengthened in the operation. The intervening lateral root between the central and cuspid, having been treated and filled, was allowed to remain. With the three crowns to serve as abutments the

Fig. 375.

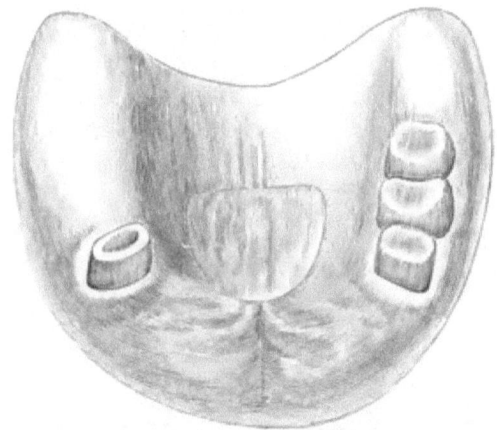

intervening spaces were bridged with artificial teeth, the extension on the left side being supported by a saddle. The spaces between the inferior cuspids and molars on both sides were bridged, the bicuspid on the left supporting the bridge, instead of the cuspid, as on the right; the left cuspid was crowned and the incisive edges of the incisors trimmed even.

Fig. 371 illustrates the case finished and shows the complete artificial restoration of the parts.

Figs. 372 and 373 represent a case in which crown- and bridge-work has been extensively applied. The occluding surfaces of the teeth were affected with abrasion. Gold crowns with porcelain fronts which presented laterally the form illustrated were mounted on the incisors and cuspids, the pulps of which were preserved. The posterior teeth were crowned with all-gold crowns. The missing teeth, except those on the inferior left side, were artificially restored with bridge-work.

In the case illustrated by Figs. 374 and 375, two superior molars and a second bicuspid on the right, and a first bicuspid and a saddle under the "dummy" representing the first molar on the left side, form the abutments. This piece, at the time of writing, has been worn satisfactorily for five years. Prior to the insertion of the bridge-work, the patient had worn artificial teeth on a plate.

CHAPTER VI.

REPAIR OF CROWN- OR BRIDGE-WORK

The fracture of a porcelain front to a permanently attached bridge is an annoying accident for both patient and dentist. It is usually attributable to failure to properly protect the incisive edge or occluding surface of the porcelain with metal, a precaution rendered necessary by the rigid character of the resistance offered the antagonizing teeth through the abutments. In most cases the porcelain can be replaced without the removal of the bridge, but the attachment is not usually so reliable as in the original piece. The following is the method usually adopted with incisors or cuspids: The surface of the backing to the porcelain is trimmed level, the platinum pins drilled out, and the holes slightly countersunk on the palatal side. A tooth similar to the one fractured, with long pins, is ground and fitted to the backing. The pins are then riveted on the palatal side, into the countersunk holes of the backing. The riveting is best done with punch forceps having a large punch, the porcelain front being protected by a piece of lead placed against the labial aspect (Fig. 376.) The heads of the pins should then be burnished smooth with a revolving burnisher. In case of the fracture of the porcelain front of a bicuspid crown or dummy, a corresponding front is selected with very long pins, and ground to fit. Holes are drilled in the gold, in proper positions, to receive the pins their full length. The

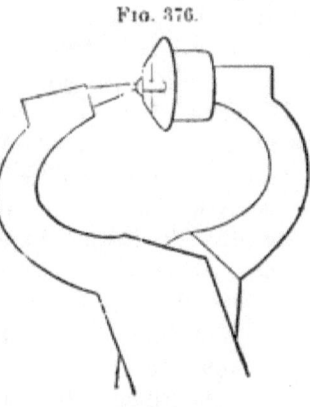

Fig. 376.

pins are then roughened a little with a sharp instrument, and pins and front are cemented to the gold. Should the holes extend through the gold at the palatal side, the cement and the ends of the pins should be covered with gold foil or amalgam. If any gold is present to protect the occluding edge of the porcelain, it should be burnished against the surface.

Bridge-work can be made with replaceable porcelain fronts which can be easily substituted in case of fracture. A description of their application will be found at page 175.

If the character of the breakage is such as will not permit its repair in the mouth, the bridge must be removed for the purpose and then reinserted. Under such circumstances the advantages of bridge-work in a detachable form are most evident.

The results following the repair of single crowns in the mouth will seldom repay the labor attending it.

Removal of Cemented Crowns and Bridges.—When it becomes necessary to remove an artificial crown, whether inserted singly or as a support for a bridge, incisors or cuspids are detached by grinding the gold or porcelain away at the palatal section over the post, which when exposed is severed, and the attachment of the cement broken up. The collar can also be slit and bent aside if found necessary. In an all-gold cap crown on a posterior tooth a hole drilled in the grinding-surface will give access to the cement, a sufficient quantity of which can then be removed to loosen the attachment; or, the collar can be divided and pried up from the root by first making a groove with a small rubber and corundum disk and then cutting the gold with a sharp instrument. With care and patience in the operation, the crowns will not usually be injured beyond repair.

When gutta-percha is the cement that has been used, a hot instrument can be placed against the metallic part of the crown, or the crown seized with the beaks of heated forceps, the gutta-percha thereby softened, and the crown removed.

CHAPTER VII.

THE HYGIENIC CONDITION OF THE MOUTH AS AFFECTED BY BRIDGE-WORK.

The probable future condition of a mouth in which a piece of bridge-work is permanently fixed is a matter of the deepest interest to both patient and operator. There is no valid reason why an artificial structure in the mouth should be more hurtful than that which is natural, provided that correct conditions are observed in its construction and proper measures are taken for their maintenance. The natural teeth demand care on the part of their owners, and all forms of artificial dentures require attention to secure their cleanliness and thus preserve the health of the adjacent tissues. What will result from the wearing of a permanently fixed bridge is almost wholly dependent, in the first place, on the proper application of principles and correct formation in every detail of construction of both crowns and bridge-work; and in the second place, upon the maintenance of cleanliness. Neglect of a single requirement will so far detract from the usefulness of the work, and may influence the final result disastrously.

Firm, properly selected abutments will not redeem incorrect conception or faulty construction; neither will the best construction remedy that which is wrong in principle or application. Self-cleansing spaces, if improperly formed, have exactly the contrary effect from what is intended, by becoming receptacles for particles of food débris, instead of preventing their accumulation.

Inaccessible spaces or interstices, which are always apt to cause uncleanliness, should be avoided. Continuity of structure of the several parts is also essential to fully insure perfect hygienic conditions.

The health of a mouth containing a piece of bridge-work constructed under these precautions can be readily maintained. For this purpose a suitably formed brush and a dentifrice are necessary. Floss silk passed through apertures around the necks of crowns, in places out of reach of the brush, and drawn along the gum under the bridge, will remove accumulations of débris otherwise inaccessible. A solution of a detergent and disinfectant mouth lotion in water, injected with a dental syringe, can be used advantageously to wash out such places and maintain a healthy condition of the gums. In addition to these measures, the crowns and bridge should be thoroughly cleansed by the dentist at regular intervals.

So cared for, a permanently fixed bridge will not militate against the absolute wholesomeness of the mouth; but it can hardly be expected that the insertion of bridge-work will insure a state of the mouth which for cleanliness will be superior to the presence of the natural teeth. Neglect on the part of the patient to perform such duties as are necessary to preserve the natural teeth in a healthy state will have about the same effect on an artificial denture. The attention required to be given to bridge-work is not greater than is commensurate with the advantages which it confers on the wearer.

CHAPTER VIII.

DETACHABLE AND REMOVABLE BRIDGE-WORK.

The evident advantages of bridge-work have stimulated the inventive genius of dental mechanists to improve the method and form of its construction and to extend its application. With these objects in view, some bridges have been made so as to be easily detachable by the dentist, and others removable by the patients themselves.

The construction of bridge-work in either of these forms overcomes the chief objections to the system. Large bridges are much more easily made in a detachable or removable form than are the smaller pieces, which present some of the best features of the permanently attached methods.

The following ingeniously constructed bridges will serve to illustrate some of the more valuable detachable and removable methods.

DR. WINDER'S SECTIONAL CROWN METHOD.

This method, an invention of Dr. R. B. Winder's, of Baltimore, presents the novel feature of constructing the crowns and forming the abutments in sections, the bridge being attached to the detachable section.

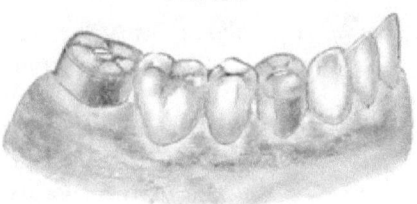

Fig. 377.

Fig. 377 illustrates a case of bridge-work made in this manner. The collar section of the artificial crown is capped and cemented on the natural crown or root, the gold forming the occluding portion of the crown, when the bridge is adjusted in position, being secured to it with a screw. The screw may be made to enter the body of the crown as in Fig. 378, or it may be

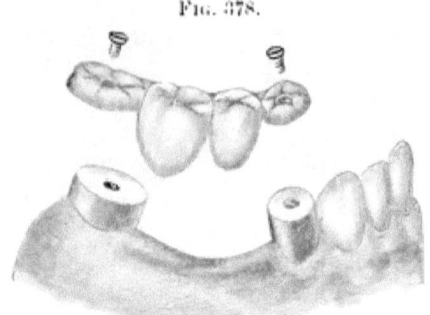

Fig. 378.

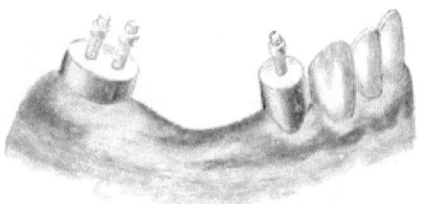

Fig. 379.

soldered to the cap on the collar, passing through the occluding section of the crown, and being secured by nuts on the screws (Fig. 379).

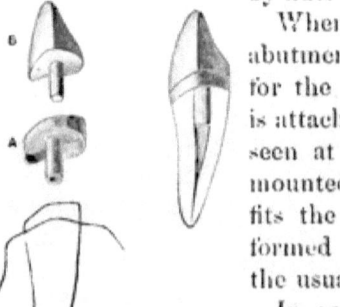

Fig. 380. Fig. 381.

When incisor or cuspid roots form the abutments, Dr. Winder first forms a cap for the end of the root, to which a tube is attached extending up the root-canal as seen at A, in Fig. 380. On this cap is mounted the crown (B), the post of which fits the tube tightly. To the crown so formed (Fig. 381) the bridge is joined in the usual manner.

In constructing a bridge of this style, the crowns forming the abutments having first been made, are removed from the mouth in a plaster impression and articulation, from which a model is made, showing the crowns in position. Each section of the bridge between the crowns is then constructed, and the crowns adjusted in the mouth. The bridges are next inserted in position, and cemented with resin and wax to the detachable sections of the crowns. The whole is then removed in investing material, in an impression-cup, or by placing the investing material in position on the bridge. After being removed from the mouth more investing material is added and the bridge and crown sections soldered together. Only the detachable sections of the crowns should be in the investment at the time of soldering.

The incisive edges can be protected and the occluding surfaces of the porcelain capped with gold as in permanently attached bridge-work, or they can be formed of the porcelain, which latter lessens the labor of construction, as the bridge is easily detached from the abutments for the purpose of repair. When the occluding surfaces of bicuspids or molars forming the bridge are to be capped with gold, the collar sections alone are first made and removed in the impression. The caps for the crowns and the bridge teeth are then formed of one continuous piece of gold plate. This is made by laying the strip of gold on a piece of lead and stamping along its length with suitable dies representing the occluding surfaces of the different teeth. The gold is then properly fitted to the collar sections on the model, conforming to the occlusion of the antagonizing teeth. The cusps are filled with solder, and the porcelain fronts, backed with platinum plate, leaving the pins straight, are fitted in position to the gold forming the caps and the backings, and cemented with wax. The porcelain fronts are next removed, without removing the backings, which are invested and soldered to the caps. When this is completed, the holes in the backings are deepened with a drill, and the pins of the porcelain fronts, having first been slightly serrated, are cemented in position with oxyphosphate. This method avoids all danger of fracture of the porcelain in soldering. Fig. 382 illustrates its application to a single crown. When the bridge is finished the root and collar section of each crown is first cemented on in position in the mouth; the surface of the detachable section

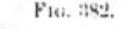

Fig. 382.

of the crown approximating the section fastened to the bridge is then heated and its surface covered with a mere film of gutta-percha. The bridge is then adjusted in position and secured by the screws or nuts. The gutta-percha prevents the secretions invading the interstices between the sections of the crown.

Fig. 383 illustrates another method, devised by Dr. J. R. Sharp, of constructing the sections of the crowns in this style of bridge-work. The part A slides in the groove B. The dovetail flange A is made of a thick piece of plate, fitted to the groove B, and riveted to a piece of platinum adapted transversely across the cap and then soldered to the removable section of the crown. Fig. 384 shows the sections of the crown in position.

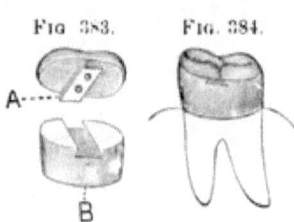

FIG. 383.   FIG. 384.

A decided advantage possessed by this method of bridge-work over others is the facility it affords in practice for the ready utilization of irregular teeth as abutments, no matter how much they converge or diverge, or lean in or out of the line of the arch.

DR. LITCH'S METHOD.

Dr. Litch's method of constructing detachable bridge-work consists in forming a shell anchorage over posts permanently fixed on cuspid roots, and anchoring the ends of the bridge with bars in slots formed in natural or artificial crowns.

Figs. 385, 386, and 387 illustrate a bridge similar to the one illustrated in Fig. 308 with this style of attachment applied. The anchorage for the cuspids is constructed as follows: The root is first capped and pivoted as for a collar crown. On the palatal portion of the collar is soldered a flange (A, Fig. 388) made of gold, No. 16 U. S. standard gauge, beveled off to the upper edge of the collar under the free edge of the gum, the object being to give a larger surface to the top of the cap. On this cap, which covers the end of the root, the anchorage post B, which is formed of iridio-platinum wire, No. 9 U. S. standard gauge, is soldered, over and back of the pin (C) which enters the root-canal, so as to allow room for the porcelain front D.

The porcelain front is ground in proper position on this cap, backed, attached with resin and wax, and removed with the cap. The cap is next invested in plaster to the edge of the collar, and a little plaster placed on the labial aspect of the porcelain front

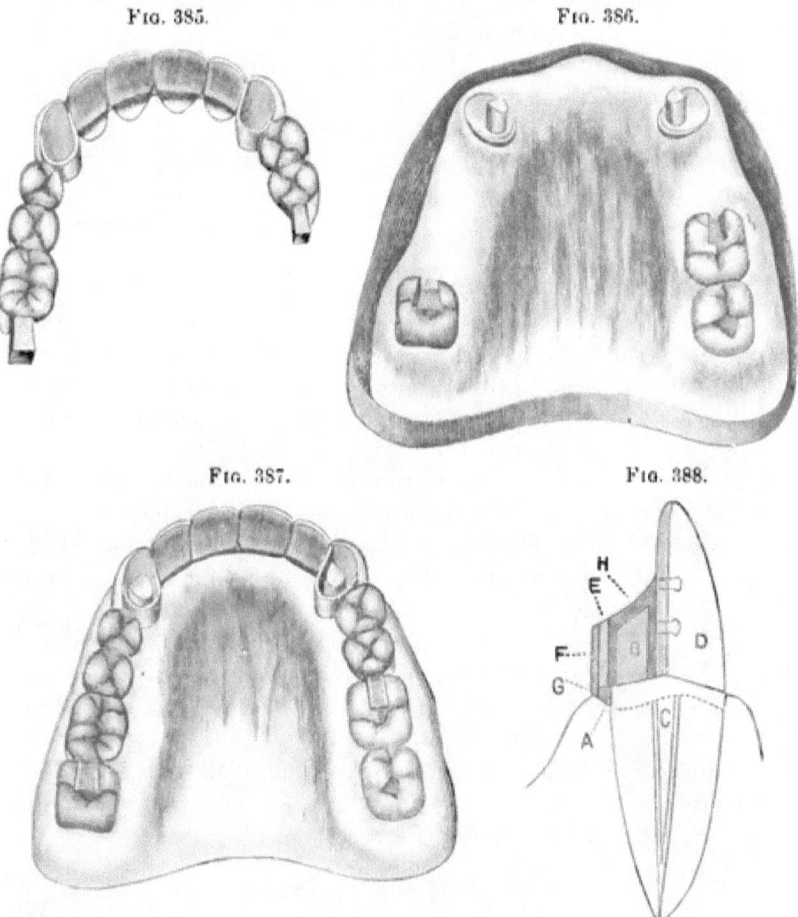

FIG. 385.   FIG. 386.

FIG. 387.   FIG. 388.

in the form of a matrix, so as to allow the porcelain to be removed and replaced. A piece of heavy iridio-platinum plate (E), No. 16 U. S. standard gauge, is then shaped into the form of a half-ring, with the ends of the plate against the backing of the porcelain front, and of sufficient size to rest on the flanged edge of the cap when completed. On the outside of this half-ring is fitted and soldered a thin strip of gold, F, of No. 28 U. S. standard gauge, which will cover the half-ring and extend beyond its cervical margin, slightly overlapping the flange of the collar (G). This forms a brace or edge on the anchorage cap as it rests on the root-cap. The half-ring is then fitted to the tooth, attached with wax, and removed with the tooth out of the plaster matrix from the root-cap, invested, and securely soldered on the inside to the backing of the tooth. The tooth and half-ring are then adjusted to the root-cap, over the post of which the ring must slide easily (Fig. 389). To this ring the bridge is soldered the same as to a crown.

Fig. 389.

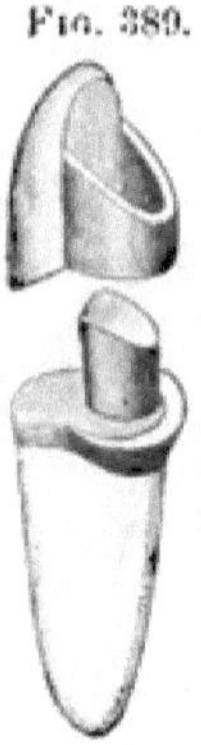

When the bridge is inserted, the cap for the root, with the post, is first cemented on with oxyphosphate. After the cement has set, the anchorage ring is filled with more cement and pressed into position upon the cap over the anchorage post. The surface of the cement (H, Fig. 388) can be protected by a metallic filling.

This form of attachment permits the bridge to be easily removed by affording access to the cement around the pin. The bar ends of the bridge are anchored in the crowns with gold or amalgam fillings, which likewise are not difficult to remove.

The anchorage cavity for a bar in a gold crown for use over a tooth with a living pulp is best made by cutting out the gold to the form of the slot required, and inserting in its place a piece of platinum of the shape of the walls of the anchorage cavity. The crown is then filled with investing material, and the metal forming the anchorage cavity soldered to the crown at the edges of the cavity.

DR. R. W. STARR'S METHODS.

Dr. R. Walter Starr, of Philadelphia, gives the following descriptions of his methods in detachable bridge-work:

"The case of Mr. W. presented difficulties of an unusual character, as may be seen by inspecting the illustration, Fig. 390, which renders detailed description unnecessary.

"It will be observed that the molars and the left second bicuspid overhang to a degree that would make the taking of an accurate impression by ordinary methods well-nigh impossible. After a careful study of the case, it was decided that two separate pieces of detachable bridge-work should be attempted, and, as an essential preliminary step, the overhanging sides of the molars and bicuspids were ground with engine corundum-wheels and points until those sides were made much less inclined, when plaster impressions were taken, first of one half, and then of

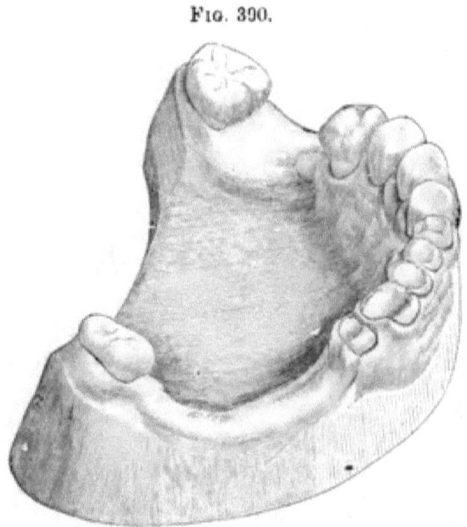

Fig. 390.

the other half, of the jaw. Gold cap crowns were closely fitted over the molars, left second bicuspid, right first bicuspid, and cuspid stump. Gold crowns were made to telescope over all the caps, which were then, by means of oxyphosphate cement, fixed firmly on the teeth. Suitable plate teeth were selected, fitted, backed, and hard-waxed in place between the telescoping crowns. After hardening the wax with cold water from a tooth-syringe, the pieces were carefully removed, invested, and soldered. The two completed bridges were easily replaced on or removed from the supporting capped teeth, and their appearance when detached is correctly shown by the illustration, Fig. 391, which also shows the capped teeth and stumps. The figure likewise shows the results of the novel method employed in crowning the incisors. Gold collars were fitted tight on the necks of the incisor stumps, and the new-style porcelain caps adjusted in the collars, and set in the oxyphosphate cement which had been packed into the collars; thus at the same time fastening the collars on the stumps and the caps in the collars, as shown completed in Figs. 391 and 392.

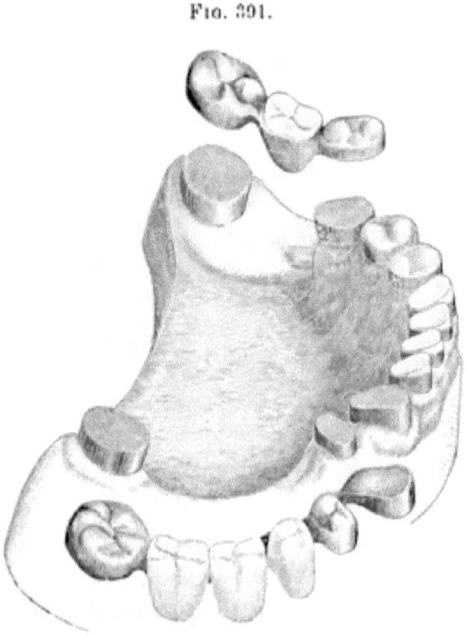

Fig. 391.

"Fig. 392 illustrates the finished crowns and bridges, which latter were secured in position by placing a small piece of gutta-percha in each of the telescoping cap crowns, which were then warmed and carefully pressed in place,—the gutta-percha filling only the spaces between the flat tops of the caps of the natural teeth and cusped caps of the bridges.

"Whenever, for repair or for any other purpose, it shall become desirable to remove one of the bridges, that may readily be done by applying a hot instrument or hot air to the caps to soften the gutta-percha sufficiently to permit the telescoping bridge to be taken off.

"A full upper vulcanite denture was made to replace the old one, which, by improper occlusion, had thrown the full force of mastication on the anterior teeth of the lower jaw, and produced

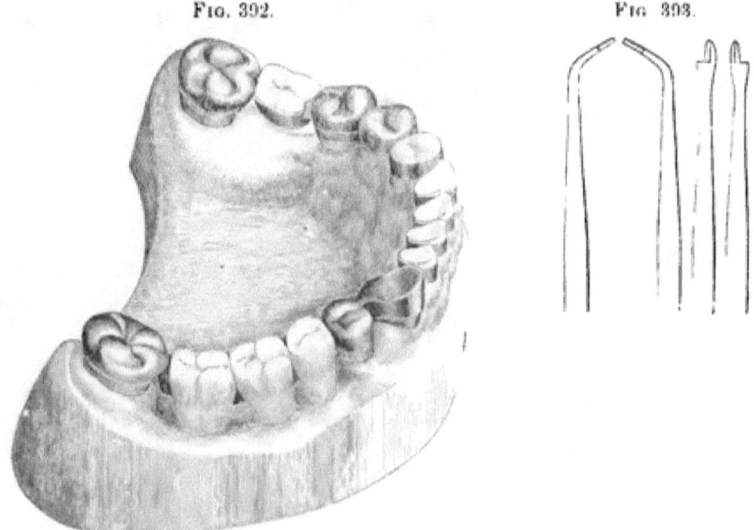

Fig. 392.   Fig. 393.

the destructive action that resulted in the deplorable loss of tooth-substance shown in Fig. 390."[1]

The next case also presented unusual difficulties. "The forward overhang of the inferior right second molar was so excessive that an impression could hardly be taken, until with corundum-wheels and points the sides of the tooth had been made parallel, or rather slightly tapering to form a truncated cone, with the neck as a base. The molar was alive and sound, but the crown was gone from the pulpless cuspid, which I suitably shaped by means of my root-trimmers (Fig. 393).

---

[1] *Dental Cosmos*, vol. xxviii, No. 1, page 17.

"An impression was then taken, the cast from which is illustrated by Fig. 394. A seamless gold collar was, by means of a slightly tapering mandrel, made to exactly fit the tapered natural molar, the lower edge of the collar cut to conform to the gingival margin; a cap piece of gold plate soldered to the top edge of the collar, and a hole drilled through the center of the completed cap (A, Fig. 394). Care was taken to so fit and proportion the cap that it would require finally pretty hard driving to send it home on the tooth; but first there was fitted to the cap a telescoping seamless collar, on which was soldered a gold plate, with cusps,

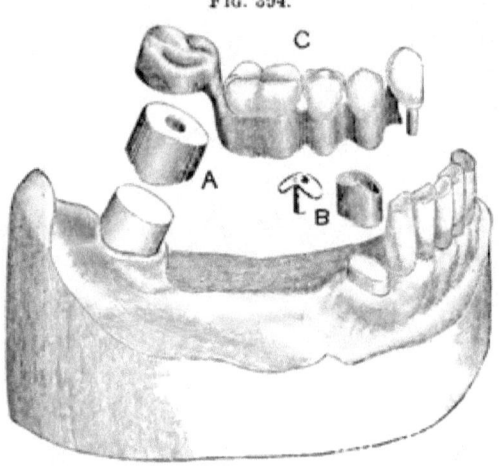

Fig. 394.

to form a molar crown as shown in Fig. 394. The molar was then thoroughly dried, slightly painted with Agate cement, and the cap, A, driven hard down with a flat pine stick held upon it and struck with a mallet; the hole in the cap enabling me to see when the cap was quite down. The cuspid was then likewise fitted with a seamless gold collar, the top edge of which was given a roof-shape, as seen above the root in Fig. 394. A piece of gold received a corresponding roof-shape, had a short section of gold tubing soldered into it, and was trimmed to the outline of the collar, beside which, B, Fig. 394, its form is seen, and to which it was subsequently soldered, after suitable investment to keep the parts in proper place. The root-canal had been previously prepared to receive the tube, which, with its roofed cap, was with stick and mallet driven hard down over the root. A piece of gold wire exactly fitting the tube had a roof-shaped piece of properly-perforated gold plate slipped over it into position on the root; became fixed in such relation by a drop of melted hard wax; was removed, invested, soldered, and finished in such shape that, excepting the hollowness, it looked like the tube and cap B.

"The relations of the occluding teeth had, of course, been determined by an articulating model, and by means of it a series of seamless gold collars and cusp-crowns were adjusted on a thin platinum plate fitted on the cast between the cuspid and second molar, and the collars soldered to the plate after investment.

FIG. 395.

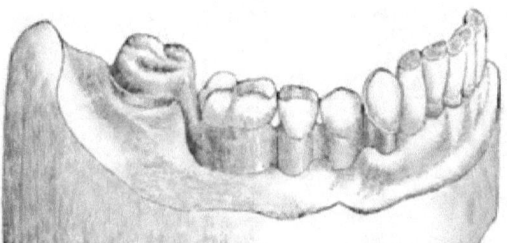

The truss thus formed received an appropriate finish by the rounding and smoothing of its basal borders. A plain plate cuspid was backed with gold plate and fitted on the roof-plate, to which, after determining its proper occlusion, it was secured by hard wax; removed, invested, and soldered. It was then put into the tube on the root; the telescoping cap put over the molar; the truss put in position in the mouth, and the whole covered with plaster and marble-dust, contained in a suitable sectional impression-tray, which enabled me to hold the mass steadily in place until the mixture was sufficiently hard to bring away cap and truss and roof-plate all in proper position. A second mixture of plaster and marble-dust, and a suitable trimming of the first mixture after all was hard, sufficed for the soldering process that resulted in the denture which, when finished, appeared as shown detached at C, Fig. 394, and mounted on the cast in Fig. 395. It went firmly to place in the mouth, and yet was removable in the possible event of accident to the denture, or for readjustment of the cusp-crowns, which latter could easily be done by warming the piece sufficiently to soften the gutta-percha, replacing the denture on its anchorages, and directing the proper closure of the occluding teeth.

DR. C. M. RICHMOND'S METHOD.

"Dr. C. M. Richmond, of New York City, in making removable dentures of the entirely soldered kind, employs a zinc die made from a cast of the anchor tooth with its cap on. He makes of crown-metal (platinum faced with gold) a collar somewhat smaller than the tooth-cap, and deep enough to reach from the gum to about a sixteenth of an inch above the cap. He then

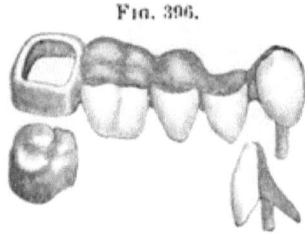

Fig. 396.

drives the die into the collar so far that the extra sixteenth of an inch can be hammered over and burnished down on the die-end to form a flanged collar. Outside of this, in the same manner, he forms another flanged collar, and then solders the two together, thus obtaining a close-fitting, stiff collar, that will not stretch in being telescoped on and off the anchorage, and is kept by the flange from being forced too far over the tooth-cap. A denture of this kind is illustrated in Fig. 396, which also shows his post and roof device in another form than that previously described."[1]

DR. PARR'S METHODS.

Detachable.—Fig. 397 illustrates a method of this style. The teeth forming the abutments lean toward each other posteriorly and anteriorly over the space to be bridged, as shown on the original model, Fig. 398. The bridge is supported by two shoulders on the abutment crowns, which slide into grooves formed in the dummies (Fig. 399). These supporting shoulders

---

[1] *Dental Cosmos*, vol. xxviii, No. 8, page 497.

DETACHABLE AND REMOVABLE BRIDGE-WORK.   201

and slots are made by shaping two pieces of plate to the form shown in Fig. 400, so that one shall telescope the other. The inner one is then invested on the outside surface and filled in with

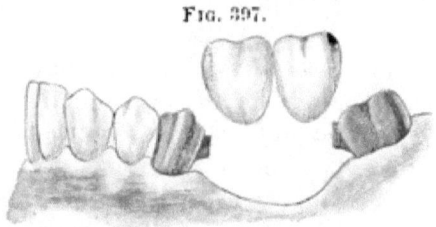
FIG. 397.

gold plate. The outer piece is then filled inside with investing material, and gold plate is flowed over the outer surface. The shoulders are first soldered to the crowns, and afterward the slots are adjusted to them and soldered in position in the bridge.

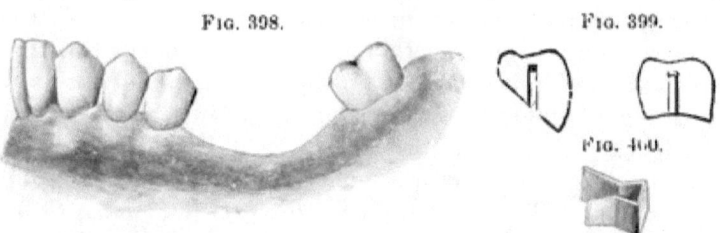
FIG. 398.   FIG. 399.
FIG. 400.

Removable Bridge.—In this style the crowns forming the abutments are permanently cemented in position, each section of the bridge between them being removable. The case illustrated in

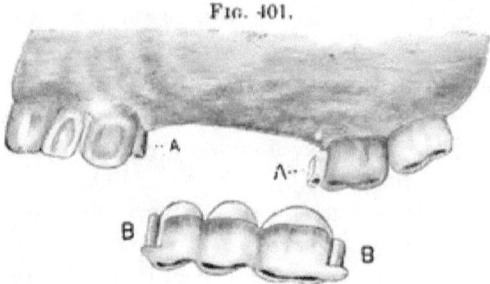
FIG. 401.

Fig. 401 will serve as a type to give the constructive details. The cuspid and molar crowns are first formed in the usual manner. A model from an impression is then made on which the crowns will be in the same position as in the mouth. A gold and platinum bar (A, Fig. 402) is then formed between the cuspid and molar. The end for the cuspid is rounded, and that for the molar flattened. This last may be done by hammering the wire flat or by soldering a piece of clasp plate transversely to it. The two ends of the bar are then fitted in sockets of platinum (B, B). The ends of the bar should be bent and the platinum sockets placed in such a position against the sides of the crowns that the bar can be easily adjusted and removed. The sockets are next soldered to the sides of the cuspid and molar crowns (A, A, Fig. 401), for which purpose the sockets and crowns should be removed and invested. The sockets are held in position when the wax is melted out by pieces of iron wire, one end of which covered with a portion of the investing material is inserted in the socket, the other end being imbedded in the investment. If preferred, the slot on the side of the molar crown can be made with a piece of platinum adapted over the flat piece of gold forming the end of the bar and then soldering the platinum to the side of the crown, the platinum being stiffened by flowing the solder over it. At this point the

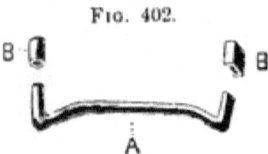

Fig. 402.

crowns and bar may be adjusted in the mouth, as well as on the model, to insure accuracy. A piece of thin platinum or gold is then perforated and slipped over the ends of the bar, which is placed in position on the crowns, and the platinum or gold adapted to the form of the attachments, and to the immediately adjacent surfaces of the crowns. These shell forms are made to assure to the ends of the bridge a perfect fit by giving them the shape of the crowns and the attachments on the crowns. To this bar the teeth constituting the bridge are fitted in their respective positions and soldered. Bending either end of the bar slightly (B, B, Fig. 401) or sawing a slit in the cuspid end of it (Fig. 410), afford the means of holding the bridge firmly in position, although it may be removed and reinserted at the option of its wearer. Fig. 403 shows the inserted bridge.

Fig. 404 shows another method of forming a socket attachment. In the figure, the socket section of the attachment is seen projecting from the side of the molar crown. The other section consists of a cap having a spring flange. The flange enters the socket, which the cap incloses on the top and sides. The spring is made by bending open a little the part of the flange marked A.

Fig. 403.

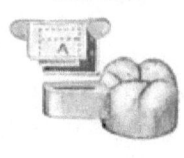
Fig. 404.

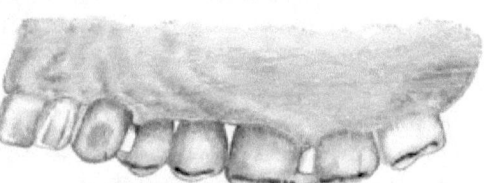

This form of attachment is constructed as follows: To make *the spring flange*, two pieces of clasp or spring gold plate about No. 23 U. S. standard gauge, one of them one-half and the other one-fourth of an inch long and from one-eighth to one-quarter of an inch wide, the exact dimensions being governed by the case in hand, are laid together, so that one end of the short piece is nearer one end of the larger piece than the other. The edge of the short piece nearest the end of the longer one is then soldered to it with a hard-flowing solder, the two being held during the operation in a blue gas flame with tweezers, and the end is trimmed square. A little whiting placed between them will prevent the solder from flowing between or joining them at the other edge. The short piece of plate is to form the spring, and is left unconnected at one end for that purpose (A, Fig. 405).

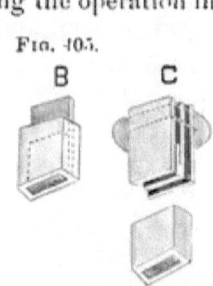
Fig. 405.

To Form the Socket.—The spring flange is first enveloped once around with a thin piece of platinum, a little deeper than the socket is required. The platinum is then enveloped with one thickness of coin gold plate, No. 32 U. S. standard gauge, about the depth the socket is to be, leaving a ledge of the platinum projecting. The platinum and gold are next removed and soldered, by holding them in a flame and using very little solder, of a hard-flowing variety, which is placed upon the ledge. The sides and ends of the socket are then filed level and the socket given a square form (B, Fig. 405).

To Form the Cap.—The spring flange having been inserted in the socket, two pieces of the spring gold plate of the same length as the socket are adjusted along its sides, the pieces being cut a little wider than the depth of the socket, so that when adjusted they shall project slightly above it. A piece of thin platinum plate is then adapted to the end of the flange, to the socket, and to the pieces of spring plate, first being perforated to allow the projecting ends of the latter to pass through it. The spring plates and the flange plate are then cemented to the platinum plate with wax, removed from the socket, invested, and soldered (C, Fig. 405).

The sockets must be so placed on the crowns that the springs at the ends of the bridge shall enter them on parallel lines. Their proper relative positions to secure this movement are readily determined by attaching the tops of the spring flanges (either temporarily or permanently) to the ends of a piece of wire of the length of the space to be bridged, which will permit the necessary adjustment. The sockets are then soldered onto the crowns.

When the bridge teeth or dummies adjoining the sockets have been fitted in position, they are withdrawn with the caps and spring flanges, and soldered to the bar, in the manner described at page 202. The gold caps forming the occluding surfaces of the bridge tooth can usually be fitted over the cap. When the socket is to be attached to the crowns lengthwise, as in Fig. 401, the spring metal plate is placed only on the labial or buccal side of the socket.

In a bridge of this style of the anterior teeth only,—where the abutments form the extremities of the piece,—the ends should be attached to the mesial sides of the crowns forming the supports; but when it also carries teeth posterior to the abutment, and the sections of the bridge are united together, the attachment should be made on the distal side, the bar supporting the anterior teeth resting in a slot formed on the palatal side of the abutment (Fig. 406). A shell crown on a cuspid can be utilized as a support for this form of attachment.

Fig. 407 illustrates a removable cuspid crown which can be used in removable bridge-work. It is constructed as follows: The end of the root is first capped. A porcelain cross-pin tooth, the pins of which are set well apart, is then ground and adjusted in position, cemented with wax to the cap, and both removed. Enough plaster to form a matrix is placed on the labial aspect of the porcelain and collar to hold them in relative position when the wax is removed. A flat post is then formed on the root-cap, extending from the labial edge forward between the pins of the tooth, the porcelain of which, between the pins, should be slightly cut out to receive it. To this is soldered the piece A which is to act as a spring. The post is formed of plate, No. 18 U. S. standard gauge, and the spring of clasp or spring gold No. 26. The post is soldered to the cap from the opposite side to the spring by investing without the tooth. The tooth is then lined with very thin platinum and with the aid of the matrix is adjusted in position on the cap, and a thin piece of platinum plate is slit at the edges and adapted over the post, cap, and edge of the collar, with its inner edge meeting the backing of the tooth. The platinum plate is then cemented with wax to the backing on the tooth, removed with it from the cap and post, and invested, the slot for the post being carefully filled with the investing material. Gold plate is then flowed over the plate and backing so as to consolidate the parts in one piece.

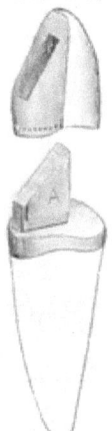

Fig. 406.

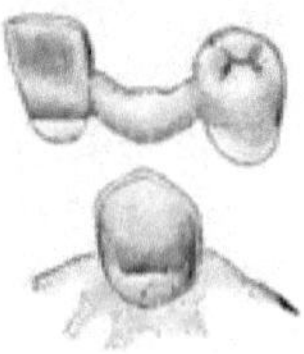

Fig. 407.

This crown when finished will fit on the cap and post, the spring of which can be bent to securely retain it. The line of the spring post on the cap as it fits in the slot in the crown should be such as will favor the insertion and removal of the bridge.

CHAPTER IX.

REMOVABLE PLATE BRIDGES.

In this style of bridge a plate is used to span the space and support the artificial teeth between the abutments. A prime requisite of removable plate bridges is that the posts and collars or any form of attachments used shall move evenly on and off the supporting roots or crowns in their adjustment and removal. To secure this, the post-cavities and crowns should be formed and the gold crowns shaped so that the lines of the center of the cavities and of the sides of the gold crowns shall be as nearly as possible parallel. To facilitate the operation, posts of wood or metal should be first accurately but loosely inserted in the root-canals, protruding a quarter of an inch, and an impression taken. On the model made from this impression the posts will be found in position as in the mouth. The plaster crowns are then trimmed to the required form. Gutta-percha or impression compound, fitted to the model and removed with the posts in position in it, can then be used to guide the operator, and gauge the preparation of teeth or roots in the mouth.

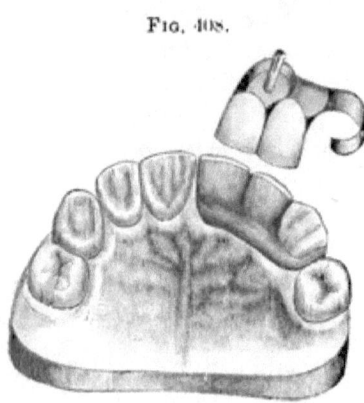

Fig. 408.

The case shown in Fig. 408 will be taken to describe the constructive details of this class of dentures. In the root of the central a tube is inserted, attached to a cap on the end of the root. Over this cap is placed an outer cap which has a split spring pivot or post fitting the tube. A narrow plate between the teeth connects the outer cap to a clasp which fits around and rests upon the cuspid.

The process of construction is as follows,—the method being similar when applied to larger dentures of this class: The root of the central is first prepared and capped the same as for a collar crown. The cap is best formed of iridio-platinum plate, No. 35 U. S. standard gauge (A, Fig. 409). A substantial piece of gold and platinum wire, from No. 16 to No. 18 U. S. standard gauge (the number being regulated by the size of the root), is slit about one-eighth of an inch so as to form a spring-post or pivot. This is easily done by placing the wire in a vise and steadily cutting it downward through the center with a saw-edged strip of very thin steel (Fig. 410). This takes only a few minutes, and is preferable to partially joining two pieces of half-round wire.

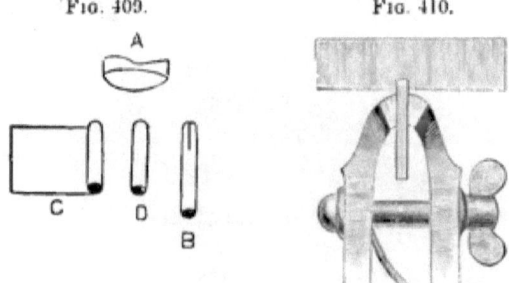

Fig. 409.   Fig. 410.

The wire is then tapped together at the slit, burnished smooth and rounded just at the end (B, Fig. 409). The tube for this pivot is formed by once encircling the pivot with a piece of iridio-platinum plate, No. 32 U. S. standard gauge, the edge of which is beveled and cut to meet the plate even and close (C). The pivot is then withdrawn, and the seam is touched along its length with the smallest possible quantity of borax. The proper manner to use borax in fine work is to grind it, mixed with water, on a slab to a cream-like consistence, and apply on the point of a camel's-hair brush. A very small piece of pure gold is placed in the seam, and the tube is held in the flame of an alcohol lamp. When a sufficient degree of heat is reached, the gold will flow along the seam and form a joint without obstructing the inside passage for the pivot. The pivot is then inserted, and the tube trimmed (D) and gauged in a gauge-plate. With a drill just the size of the tube the root-canal is enlarged so that the tube will fit in tightly. This plan prevents weakening of the root by too great enlargement of the canal. A hole the size of the tube is made through the cap, and cap and tube are then adjusted (Fig. 411), and the pivot being withdrawn, they are removed, invested, and soldered (A, Fig. 412). The cuspid, which because of its conical formation is one of the most difficult teeth in the mouth to clasp, is trimmed sufficiently to partially square its approximal sides, and the palatal portion is notched slightly

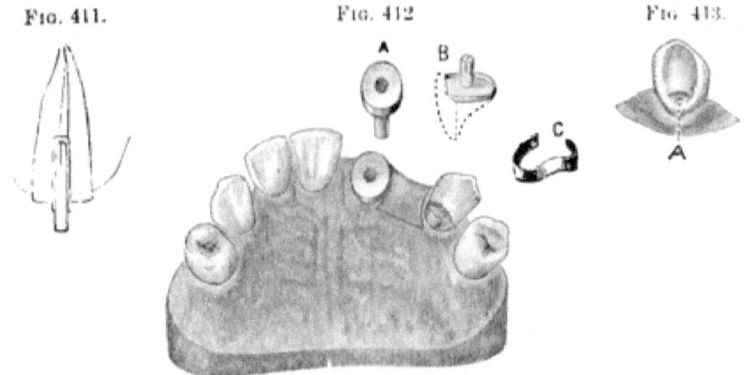

Fig. 411.   Fig. 412.   Fig. 413.

(A, Fig. 413), to form a shoulder for the clasp to rest upon. This notch can be safely made, as the enamel is very dense at the point indicated.

A gold plate is swaged to fill the space between the central root and the cuspid as shown in Fig. 412. The cap and pivot are adjusted on the central root and the plate is then fitted in the mouth, pressed tightly in position against the gum, and retained there with a little wax, which is cemented to the plate and pressed against the cuspid and side of the cap. An impression of the parts and an articulation are then taken in plaster. The cap, pivot, and plate being removed in the impression, they will be presented on the model made from it in exactly the same position as in the mouth. A second or outer cap is then constructed for the root-cap, the band being made very narrow at the approximal and palatal sides, and open at the labial side, as the porcelain tooth to be attached will serve in its stead (B). The pin is then soldered fast in the outer cap, and a clasp of clasp gold (C), No. 23 to No. 24 U. S. standard gauge, is formed to extend well around the posterior approximal portion of the cuspid. The outer cap having been placed in position on the inner one, the plate extending from the central to the cuspid is cemented to it and to the clasp with wax, removed, invested, and the parts soldered together. Aided by the plaster articulation, teeth are ground and fitted by the model, backed, attached to the plate with wax, and inserted in the mouth. Platinum foil is then burnished to the form of the notch on the cuspid, the clasp fitted over it and attached with wax, removed with the plate, and soldered to it simultaneously with the teeth.

Fig. 414.

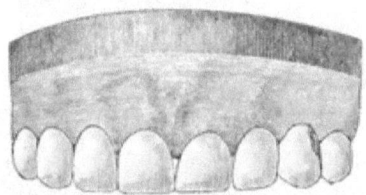

When the piece is finished and fitted in the mouth, the inner cap—the end of the tube having been closed with gutta-percha—is first cemented on the central root. A very small quantity of oxyphosphate is used, and while it is yet soft the plate is adjusted in position, and allowed to remain there until the cement has set. The split pivot is sprung open a little and forced to place. With the aid of the clasp around the cuspid it will be found to perfectly secure the plate. Fig. 414 shows the bridge in position. If the plate forming the bridge is tightly adjusted against the soft tissues, and removed in that position in the impression taken with the caps, it will always be found to fit in a similar manner when the bridge is finished. Should the clasp cause decay or abrasion of the cuspid, the tooth can be excavated to a slight depth under the clasp, and filled with gold. This is best done by making a few retaining-pits, filling them with a hand-plugger, and then inserting the main body of the gold in the ordinary manner, the Herbst method being useful in condensing the foil. Such a filling inserted at any time will prevent injury from a clasp. A denture of this style can be made to pass intervening teeth.

In the artificial replacement of the lower teeth in a case such as is illustrated in Fig. 415, a plate bridge possesses many advantages. In the construction of such a denture, the teeth are first properly shaped. Gold crowns (Fig. 416), with sides as nearly as possible parallel the one with the other, are then made and fitted to the bicuspid and molar. This operation is much facilitated by shaping the external surface of the crown with metal. On the crowns, at the buccal sides, a narrow shoulder (A) is constructed to sustain the collars and bridge in position. In some

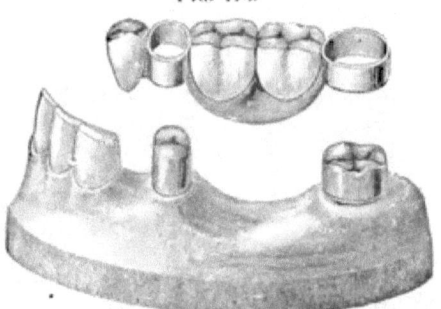

FIG. 415.

cases this shoulder is placed on the approximal side to better advantage. The crowns are then adjusted in the mouth, a small quantity of wax being applied inside of the crowns when necessary to retain them in correct position. A piece of plate is next swaged and fitted between the crowns and attached with wax as described in the previous case. An impression and articulation of that part of the mouth are then taken with plaster and the crowns and plate removed in it. On the model made from this impression the crowns and plate will appear in position. Collars reaching from the cervical to the occluding edge are fitted to these crowns. They are made by first forming a collar of ample width of thin platinum, about No. 32 to No. 34 U. S. standard gauge, which is easily adapted to the form of the crown, and on the outside of this fitting a slightly narrower strip of gold clasp plate, about No. 30 U. S. standard gauge. The gold and platinum are then cemented with wax, removed, invested, and soldered together with gold solder. A perfect-fitting and reliable collar is thus formed (B, Fig. 416).

The collars, though fitting accurately, should move easily over the crowns, as they can be readily tightened when the case is finished. If a collar is troublesome to adjust and remove, cut the side opposite to the one attached to the plate, and spring it open a little. After fitting the teeth it can be again united when they are being soldered. This collar and shoulder form a support preferable to a partial or an entire double cap, being less difficult to keep clean. A collar is more easily constructed, and

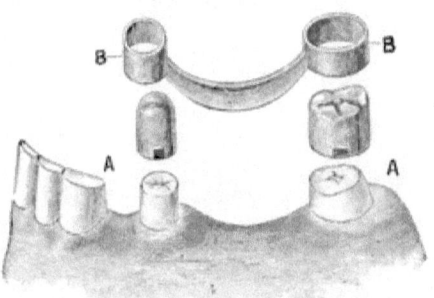

Fig. 416.

also permits the position of the bridge to be altered by the removal of a little of the shoulder or of the upper edge of the collar, and is a secure but less rigid attachment than other methods.

The collars and plate are next cemented with wax, removed, invested, and strongly soldered together (B, B, Fig. 416). The artificial teeth are fitted in proper position on the plate by the aid of a plaster articulation, attached with wax, and, if preferred, adjusted in the mouth without the crowns. The bridge is then invested and finished. The attachment of the artificial teeth to the plate can be of either gold or rubber. Whichever is adopted, the first bicuspid is best supported by being soldered to the collar. If iridio-platinum is used in the construction instead of gold plate, and the soldering done with pure gold, porcelain body can be used. When ready to be inserted, the crowns are first adjusted with cement, and then the bridge, which is left in position until the cement sets. By burnishing the collars they can be made to clasp the crowns as firmly as desired. Fig. 417 shows the denture in position.

Fig. 417.

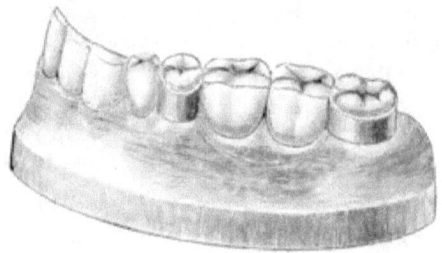

Figs. 418 and 419 represent an upper removable plate bridge. In its construction the cuspid roots were first capped, tubed, and pivoted, and the molars crowned with shoulders formed on the

Fig. 418.

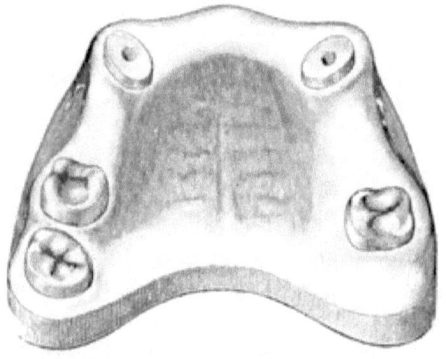

Fig. 419.

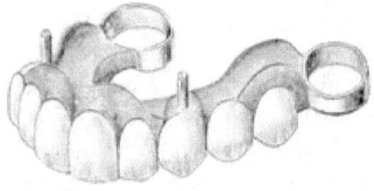

buccal sides. The plate intended to connect the abutments was then adjusted in position as has been described. An impression was next taken and a model and articulation made. The cuspids were then double-capped and collars formed on the molar crowns. The double caps, pivots, collars, and plate were next soldered together. The artificial teeth were attached with vulcanite, the gum section being formed with pink. In order to avoid any warping, which might readily occur in the construction of so large a denture as this, the plate may at first be swaged up, as in ordinary artificial dentures, to cover the whole of the hard palate. A shallow groove can be made around the palatal surfaces of the teeth, and after the final soldering the plate can be cut along the line of this groove, the portion covering the palate being removed. The groove will insure a close fit for the palatal edge of the plate.

FIG. 420.

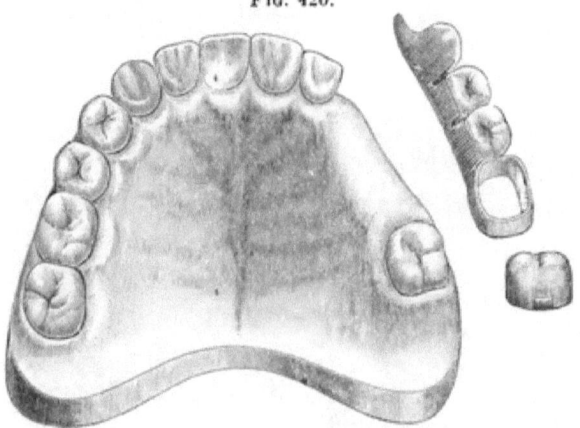

In the case represented in Fig. 420 the natural teeth were very short. The posterior approximal side of the molar was decayed to such an extent that the pulp was nearly exposed, and considerable irritation of the investing gum-tissue had been caused by the clasp of a plate worn by the patient working upward against it. The patient declined to have a plate made which would extend across the palate. The lateral was hardly strong enough to support a permanent bridge. The molar was capped, and a removable appliance constructed with a band which slipped over the cap and rested on a shoulder on the mesial side to form the posterior abutment. The lateral was notched and clasped for the anterior support. Figs. 421 and 422 are two views of the appliance in position.

Fig. 421.

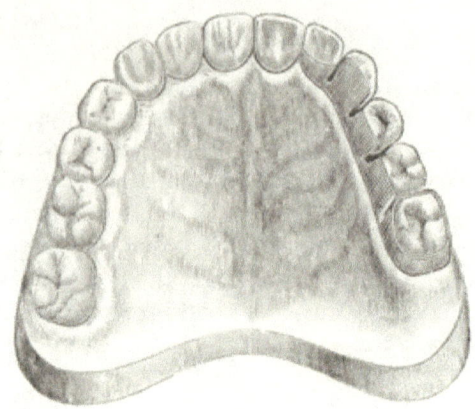

Fig. 422.

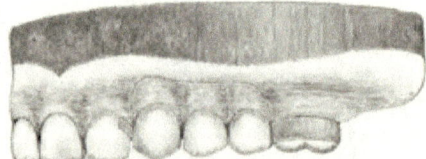

Fig. 423.

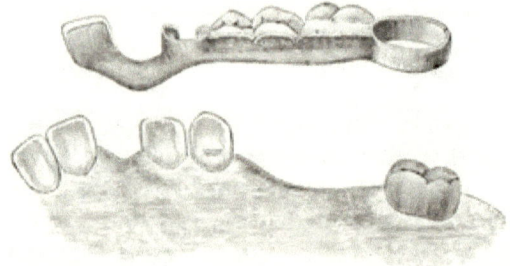

Fig. 424.

Fig. 423 represents a case in which a central incisor is replaced on an extension of the plate, the cuspid being partly encircled by a clasp with a shoulder resting on the palatal section of the tooth. Fig. 424 shows the denture in position.

Fig. 425 represents a case in which a bridge-plate was inserted without crowning either of the abutments. The clasp of a plate which had been worn for some years had worked upward and abraded the distal section of the cuspid to such an extent as to expose a large portion of the root and superinduce decay.

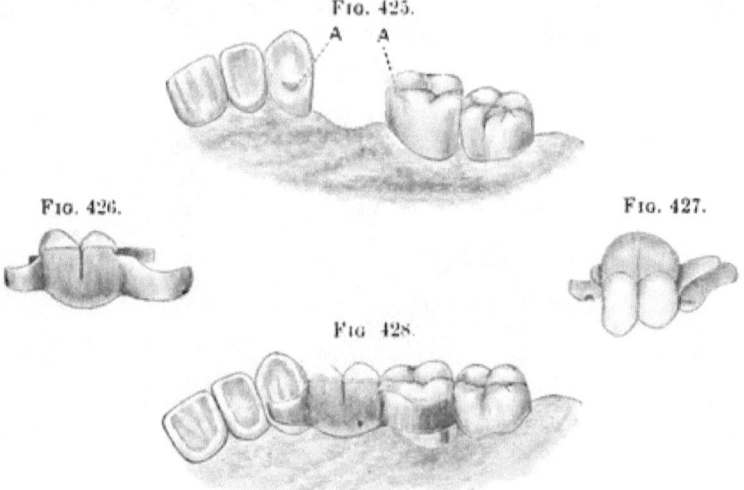

Fig. 425.

Fig. 426.   Fig. 427.

Fig. 428.

The cavity was filled with gold, and the gingival border, by treatment, brought nearly to its normal position on the tooth. A plate bridge, such as is represented in Figs. 426 and 427, was then constructed. Clasps, with flanges resting on little shoulders formed at A, A, Fig. 425, support and retain it. A flange such as was used in this case is best made subsequent to the construction of the plate and clasps, by burnishing a piece of platinum foil in the mouth to the form of the shoulder and the side of the tooth upon which it is to rest, adjusting the clasp over it, and cementing with wax, then removing, investing, and soldering. Wherever the platinum is placed the solder will flow and fill all the space between it and the clasp. This gives the clasp the exact form of the tooth.

Fig. 428 shows the denture finished and in position. If the teeth are dense in structure, an attachment of metal held in proper position against the lower portion of a crown will be worn a long while before it causes injury to the parts. Filling, or crowning, if necessary, can subsequently be resorted to.

By a correct application of the methods just explained and illustrated, a piece of removable bridge-work of this style can be devised for many cases.

Some forms of attachment described in the chapter on "Removable Bridge-Work," such as Dr. Parr's, can also be used in combination with this style of dentures.

FIG. 429.

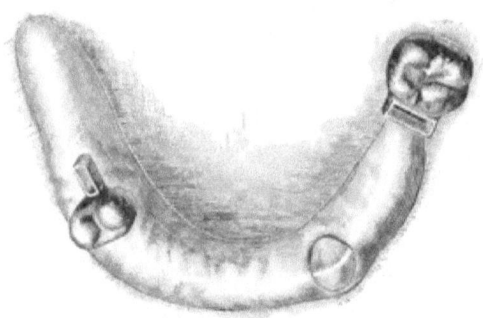

FIG. 430.

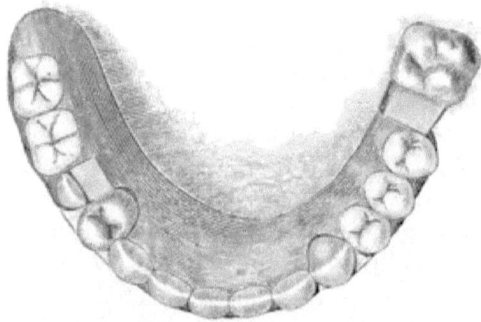

Fig. 429 shows the cast of a lower jaw in which only the left second molar, left cuspid, and right first bicuspid remained. The molar and bicuspid were fitted with gold cap crowns, and spring socket attachments (Dr. Parr's form) were soldered in proper positions on the crowns, as illustrated. The completed denture in position supported by the attachments is seen in Fig. 430. The under side is shown in Fig. 431.

FIG. 431.

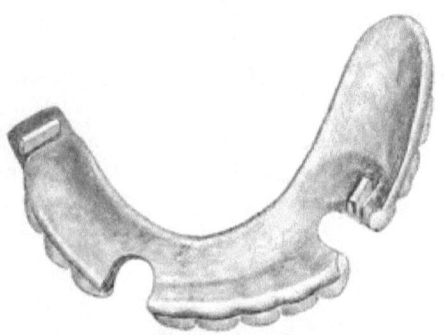

Fig. 432 represents the articulated cast of a case in which a similar form of attachment and a clasp were used. This is illustrated in Fig. 433, which needs no description.

FIG. 432.   FIG. 433.

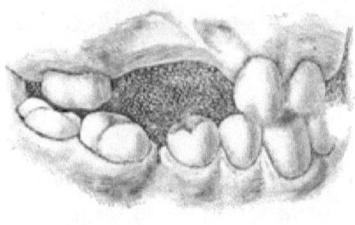 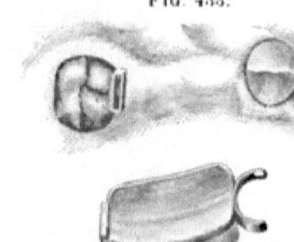

FIG. 434.

Fig. 434 shows the denture in place. It was constructed of vulcanite, and made for and placed in the mouth of a patient exhibited at a clinic of the Odontological Society of Pennsylvania, at Philadelphia, in December, 1888.[1]

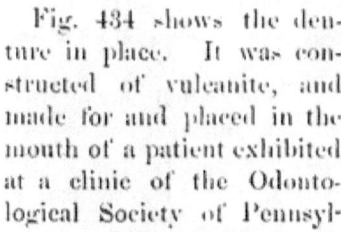 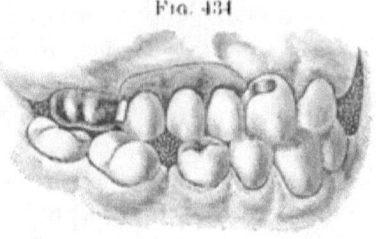

---

[1] *Dental Cosmos*, March, 1889.

DR. WATERS'S METHODS.[1]

Dr. T. S. Waters, of Baltimore, explains the advantages of his removable plate bridge-work as follows: "In the system I present the denture is retained securely and steadily in the mouth, yet is readily removed and replaced at pleasure by the wearer. The pressure and strain are distributed properly over all the structures and tissues available for the purpose, and the roots and crowns to which the denture is attached are so prepared that there is no place for the lodgment and retention of food, and when the denture is removed, both it and the mouth can be thoroughly cleansed. Should the roots or other tissues be attacked by disease, thus requiring treatment, or should repairs

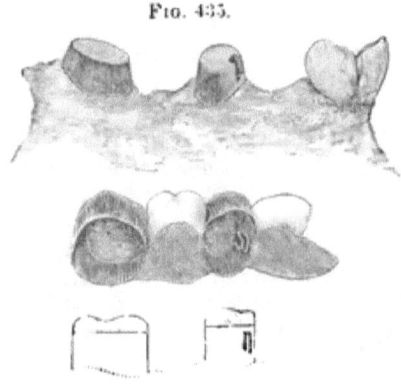

Fig. 435.

to the mechanism become necessary, the removable bridge-work offers facilities for those purposes not to be found in permanent dentures."

Dr. Waters thus describes the formation, combination, and application of his devices to cases of removable plate bridge-work:

"My devices are three in number, each of which may be used alone, or two of them or all three may be combined and applied in the same case, as the position, character, and relation of the teeth and roots remaining in the mouth may seem to indicate.

"The first is a gold crown fitted to and sliding on a cap

---

[1] *International Dental Journal*, April, 1889, p. 197.

attached permanently to the root or natural crown. This cap is made high and has on one side a longitudinal groove. The gold crown has soldered on the inside a spring catch, which works in the groove on the cap, and holds the crown firmly in its place. The character of the device is seen in application to the case represented in Fig. 435. Fig. 436 shows the denture in position.

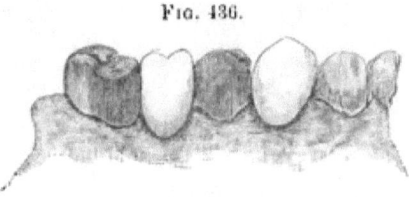

Fig. 436.

Fig. 437 gives the lingual aspect. It will be readily seen that under proper circumstances two or more roots or teeth may be fitted with this device, the gold crowns being soldered to and made a part of the denture, making the whole a piece of bridge-work capable of being removed, cleaned, and replaced at will. The spring catch regulates the firmness of retention.[1]

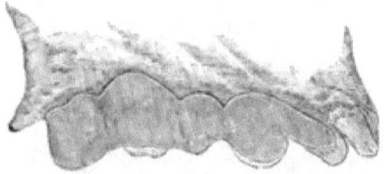

Fig. 437.

"The next device is the box cap and split post; the box cap being fitted permanently to the root, and the split post being soldered to the plate bearing the teeth. The box cap is the usual cap, with a box or tube soldered to it and extending into the root, the cap end of the tube being open. The split post is so secured to the denture as to slide snugly into this tube, the firmness of retention being regulated by pressing the split slightly open when necessary. This device, like the first, may, under

---

[1] Dr. Waters has patented this invention and donated it to the profession.

proper circumstances, be used by itself in any case, as shown in application in Figs. 438, 439, and 440, in which the whole denture is supported by box caps and split posts adjusted to the roots of the six anterior teeth.

Fig. 438.

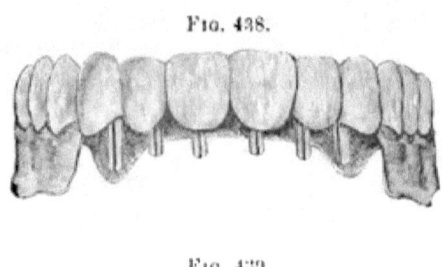

Fig. 439.

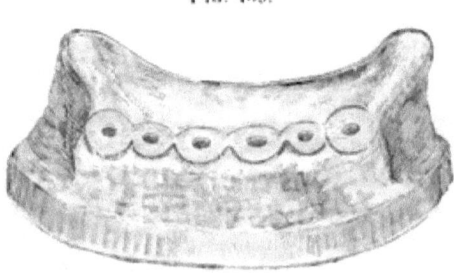

Fig. 440.

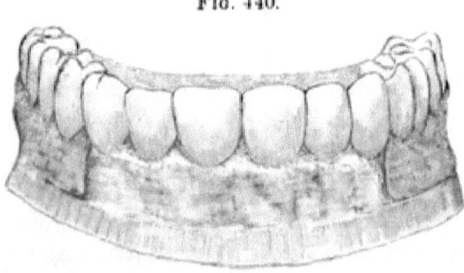

"The third device consists in soldering to the side of the gold crown covering the natural tooth a split pin or post, which is inserted into the open tube attached to the denture.

"As before remarked, these devices may be used singly or in combination in any one case. In one of the dentures illustrated the box cap and split post alone are used; in another, the cap, gold crown, and spring catch are used; in the case illustrated in Figs. 441, 442, and 443 the three are applied, in which the entire denture is attached to and retained by two cuspids, a bicuspid,

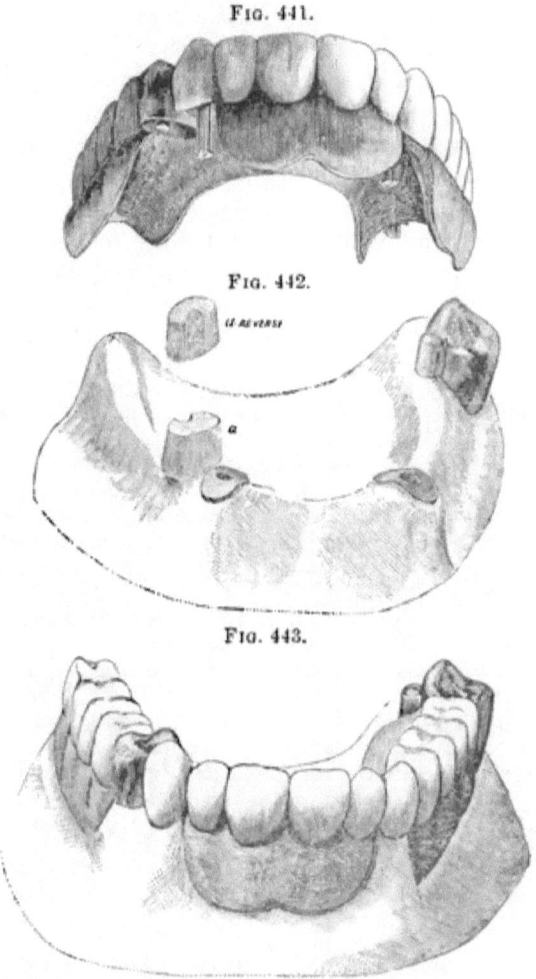

Fig. 441.

Fig. 442.

Fig. 443.

and a molar. In all this, great care must of course be taken in the preparation of the roots and natural crowns, to protect them against the action of destructive agents."

CHAPTER X.

THE LOW BRIDGE.

Dr. J. E. Low, of Chicago, is the reputed inventor of bridge-work formed with self-cleansing spaces and supported by means of cemented collars or collar crowns.[1]

The following is Dr. Low's method of using his step-plug crown (see page 126) in conjunction with all-gold cap crowns in bridge-work. Fig. 444 shows a case with the prepared molar crowns and cuspid roots. The bridge constructed according to this method is seen in Fig. 445, and adjusted in position in Fig. 446. Figs. 447, 448, and 449 illustrate a case of four incisors supported by two step-plug crowns on the cuspids.

Dr. Low gives the following instructions relating to the construction of shell crowns or anchorages on cuspids, to support a bridge of the four incisors in a case such as is shown in Fig. 450: "I first measure the tooth with strips of tin, and make the gold

---

[1] The construction of bridge-work supported by collars or any form of collar crowns cemented on teeth or roots, according to a recent judicial decision, is not at present free to public use. The two claims which reserve the use of these methods to the inventor as specified in the letters patent granted, are as follows:

"What I now claim as new is: 1. The herein-described method of inserting and supporting artificial teeth, which consists in attaching said artificial teeth to continuous bands fitted and cemented to the adjoining permanent teeth, whereby said artificial teeth are supported by said permanent teeth without dependence upon the gum beneath.

"2. An artificial tooth cut away at the back, so as not to present any contact with the gum except along its front lower edge, and supported by rigid attachment to one or more adjoining permanent teeth, substantially as and for the purpose set forth."

A denture between two or more teeth or roots, supported by such means as bars extending from it anchored into teeth or caps or crowns which do not encircle the teeth or roots, with the artificial teeth resting on or pressing into the gums and not formed with what are termed and described as self-cleansing spaces, and removable bridge-work, are not, in the opinion of experts, included in the meaning or specifications of this patent.—G. E.

bands and cut out the outside lower portion of the band before beginning to fit. In fitting, as the band is being driven down, cut away any of the band that touches the gum before all touches; never drive the band under the gum, as inflammation will prob-

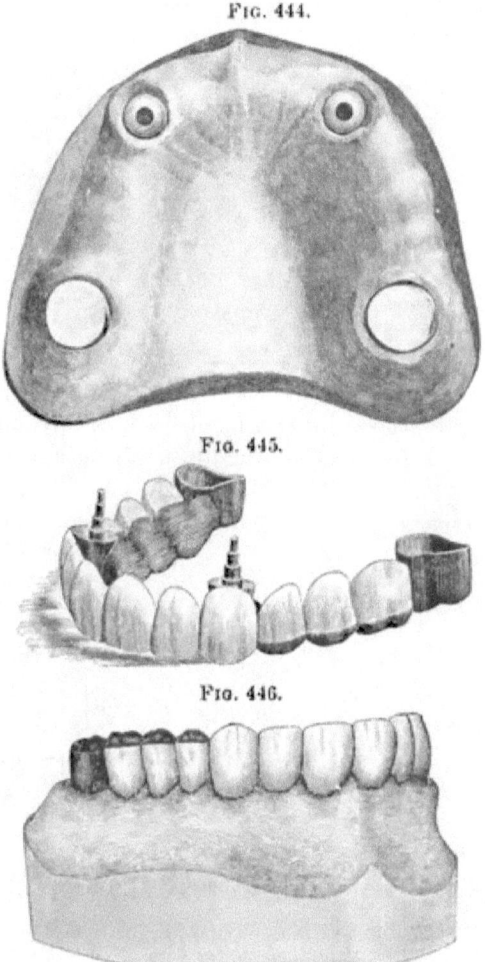

Fig. 444.

Fig. 445.

Fig. 446.

ably follow. I mention this, as I have seen many attempts to get rid of the bands by driving up under the gums and cutting them out on the front, until they are too narrow for strength. The bands should be heavy and strong, and the patient should understand that if he expects to get rid of the annoyance of the plate he must sacrifice his dislike to showing gold. After driving the bands up close to the margin of the gums, as the cuspid teeth are very tapering, the bands will have to be taken in at the

Fig. 447.

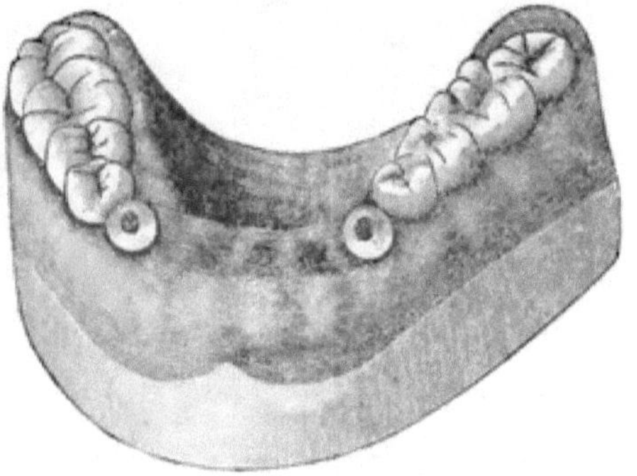

bottom. To do this I slit the band about a third of its length up, then place it on the tooth again, lap it over to bring it to a close fit, and then take it off and solder. Continue taking it in wherever it does not perfectly fit the tooth, and after a good fit

Fig. 448.　　　　　　　　　　Fig. 449.

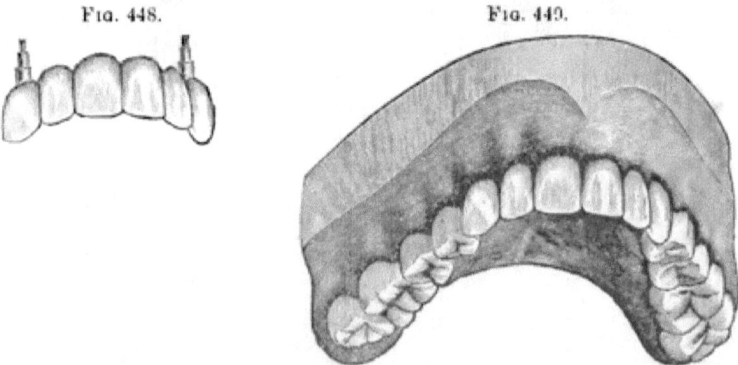

is obtained proceed with the construction of the bridge by taking an impression and articulation.

"In adjusting the bridge when finished, first try it on to see that it fits and that the articulation is all right. Fig. 451 shows the case ready for adjustment. Next dry the teeth upon which the bands are going, and then mix your cement. This should be mixed to about the consistence of thick cream. It must be neither too thick nor too thin, or the adhesion will not be strong enough to hold. Cover your teeth with cement, and then the

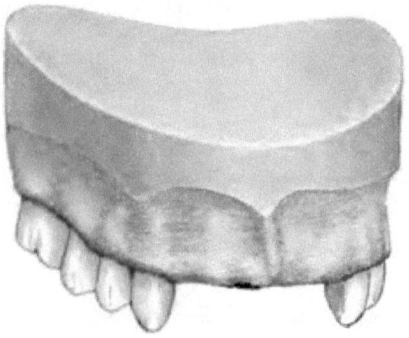

Fig. 450.

inside of the bands. Place these on the teeth and carefully mallet up into position. For this purpose I use a steel instrument with a crease or groove in the end. The teeth must be kept dry after the case is in position until the cement is well set.

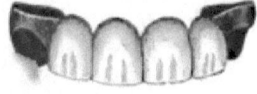

Fig. 451.

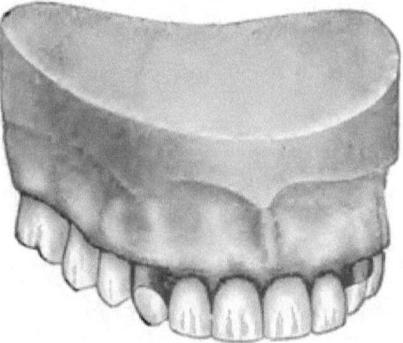

Fig. 452.

After this is done bevel the edges of the bands and burnish close to the teeth, and if properly done they will resemble gold fillings. In Fig. 452 we have the case completed. I always impress upon my patients the necessity of having the case reset immediately should it become loose, and advise them to have their cases examined at least once a year."

Dr. Low has invented for use in crown- or bridge-work a ready-made metallic socket or shell, into which fits a corresponding porcelain front, which can be replaced in case of fracture.

Fig. 453.   Fig. 454.   Fig. 455.   Fig. 456.   Fig. 457.

Fig. 453 shows a socket shell. Figs. 454 and 455 represent a socket with the porcelain in position. Fig. 456 represents the socket as made for the incisors and cuspids, and Fig. 457 the porcelain in position.

CHAPTER XI.

DR. KNAPP'S METHODS.

Dr. J. Rollo Knapp, of New Orleans, has introduced some novel methods in crown- and bridge-work, for effecting artistic results and continuity of structure.

In crown-work, Dr. Knapp invests for soldering so that the parts to be united and filled form a miniature mold. Into this mold, at a high heat, with a pointed flame from his blow-pipe, he flows solder, which fills the interstices, joins the parts, and assumes the form of the mold. The following is a brief description of his methods:

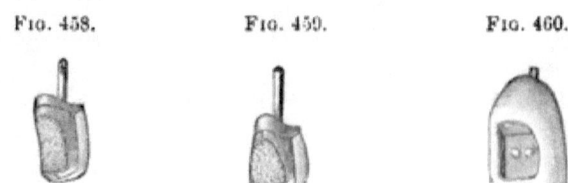

FIG. 458.   FIG. 459.   FIG. 460.

In constructing a gold collar crown with porcelain front, for an incisor or cuspid, 22-carat The S. S. White Mfg. Co.'s collar gold, No. 28 standard gauge, is used for the collar, which is formed by adapting the gold to the root by the aid of pliers. The cap to the band is then made of pure gold, No. 34 gauge, and a gold pin soldered in position for the root-canal. A plate tooth is then ground in proper position, backed with pure gold, and fastened to the cap with wax. On being removed from the mouth after proper adjustment, the side and incisive portions of the wax, including the edges of the backing and contiguous portions of the cap, are completely enveloped with pieces of pure gold No. 34 gauge (Figs. 458 and 459). The crown is then invested so that when the wax is removed the backing on the tooth with the gold on the sides shall form a small mold or pocket (Figs. 460 and 461). When the investment is heated, the flame of his blow-pipe is played over its surface until the mass is aglow, when the point of the flame is thrown into the mold by rapid thrusts until the solder melts like wax and fills every part of the mold with liquid gold. This gives an excess of gold which affords ample facilities for contouring in the process of finishing (Figs. 462 and 463).

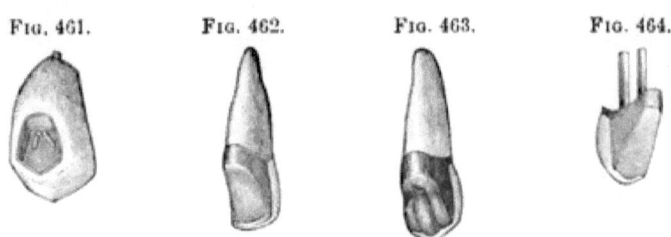

FIG. 461. FIG. 462. FIG. 463. FIG. 464.

In constructing a bicuspid crown with porcelain face, the natural tooth is ground down to the gingival edge and capped similar to a cuspid. A bicuspid porcelain front is then ground and fitted in position (Fig. 464), and the remaining portion of the crown is shaped in wax to the form required. A die of the grinding-surface is then made in metal, a cap stamped in pure

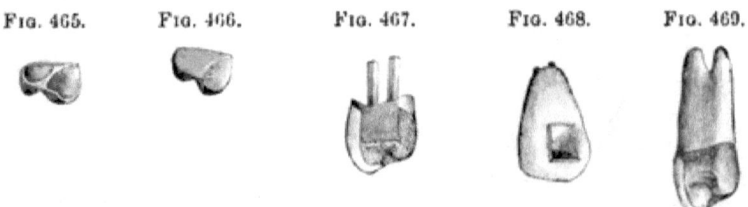

FIG. 465. FIG. 466. FIG. 467. FIG. 468. FIG. 469.

gold, No. 34 U. S. standard gauge (Fig. 465), and the cusps filled with 20-carat gold solder. This cap is next trimmed (Fig. 466) so as to fit when placed in proper position against the end of the porcelain cusp, for which purpose sufficient wax must be removed. The approximal surfaces are enveloped and the palatal portion of the collar protected with pieces of pure gold,

No. 34 standard gauge, which are slit to facilitate their adjustment (Fig. 467). This leaves the palatal portion open when the crown is invested and the wax removed, which last should be done with hot water. Fig. 468 shows the invested crown ready for soldering, in which operation the parts are filled in and joined with 20-carat gold solder. The result when finished is a solid gold crown with a porcelain front (Fig. 469).

All-gold bicuspid and molar crowns are formed by Dr. Knapp in a similar manner, but as porcelain fronts are not used in these cases, the thin gold plate is placed entirely around the labial aspect of the wax model. In crown-work, after the cap has been made, an impression is generally taken and a plaster

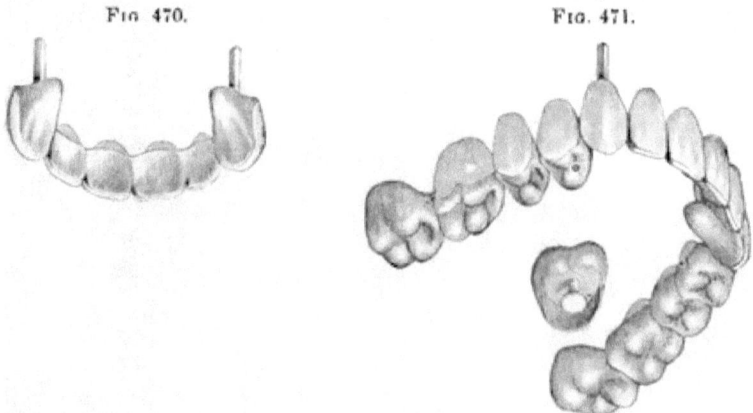

Fig. 470.   Fig. 471.

model and articulation made to facilitate the subsequent construction.

Dr. Knapp's method of constructing bridge-work is similar to that in general use except in the investing and the soldering of the parts together, the latter being done with the blow-pipe in a manner similar to that above described.

Figs. 470 and 471 illustrate specimens.

Dr. Knapp's compound blow-pipe (Fig. 472) consists of a miniature blow-pipe in which the ordinary illuminating gas (carburetted hydrogen or coal-gas) flame is combined with a current of nitrous oxide from a cylinder of the condensed gas. The combination of these gases in combustion forms essentially a carbo-oxyhydrogen flame.[1] By means of a yoke and set-screw, the valve of the cylinder is connected with the tubes and valves of the blow-pipe, so that the proportions of the mixture

Fig. 472.

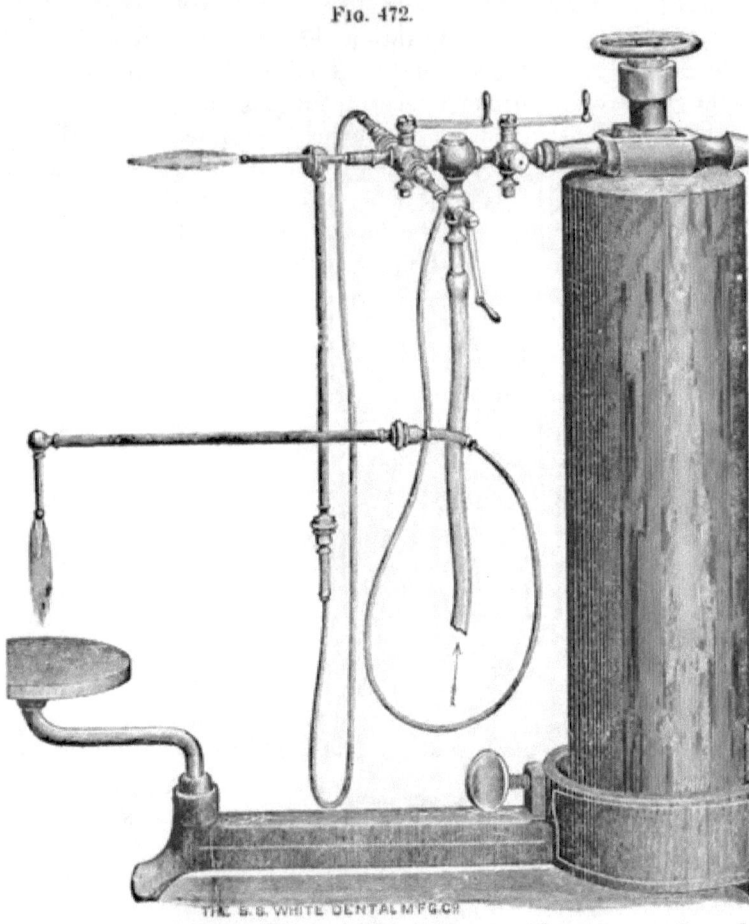

[1] The ordinary compound oxyhydrogen blow-pipe flame is produced by
  2 volumes of hydrogen ; 1 volume of oxygen.
Carburetted hydrogen consists of
  2 volumes of hydrogen ; 1 volume of carbon.
And nitrous oxide of
  2 volumes of hydrogen ; 1 volume of oxygen.
Consequently Knapp's blow-pipe flame is produced by a mechanical mixture of
  2 volumes of hydrogen ; 1 volume of oxygen ; 1 volume of carbon.

of nitrous oxide and the illuminating gases are under perfect control. The flame-jet can be diminished to half an inch in length, and at that size will melt a small piece of gold plate.

This blow-pipe is useful for many purposes in the laboratory of the present time, especially in forming solid gold backings to dummies for bridge-work, strengthening seamless gold crowns, and forming solid gold crowns.

Dr. Knapp has exhibited to the profession some very fine specimens of crown- and bridge-work, and presented much that is novel and interesting, as well as encouraging to the artistic element of prosthetic dentistry. The real value of processes or methods, however, depends on their practicability. Therefore, in contemplating that which is novel and beautiful in connection with dental art, we must be governed by this fact in estimating its value. Judged from this stand-point, Dr. Knapp's special methods, while admitting of the highest artistic results, embrace some processes which, on account of their intricacy, are not likely to be generally adopted in practice.

CHAPTER XII.

DR. MELOTTE'S METHOD.

Dr. G. W. Melotte, of Ithaca, N. Y., describes the construction of a bridge supported by a gold crown, and a shell or partial gold crown, and the use of his invention, "moldine," in connection with fusible metal in crown- and bridge-work, as follows:[1]

"Fig. 473 illustrates a case for the supply of a lateral and a bicuspid. In this instance the cuspid could be cut off, and the root collared and capped in combination with a pin entering the enlarged root-canal; but as there may be grounds for objection to cutting off sound teeth, I obviate the necessity by cutting a shoulder on the lingual portion of the cuspid, and suitably shaping its sides to permit a close fitting of the collar just under the free margin of the gum. A narrow strip of pure pattern tin, bent tight around the tooth-neck, and cut through with a knife at the lap on the labial surface, will serve as a measure for the length of a strip of 22-carat gold plate, No. 29 thick, and as wide as the length of the distal side of the cuspid. The ends of the gold are then squared, and with round-nosed pliers brought evenly together, to be held in flush contact by the soldering-clamp as shown in Fig. 474. The soldered collar, with its joint side inward, is then adjusted on the tooth as accurately as possible, giving slight blows with a mallet until the collar touches the gum, when it should be marked to indicate the necessary trimming to conform it to the gum contour. After it has been thus trimmed, the edges beveled, the labial part swelled with contouring pliers, and the lingual part cut down to about one-

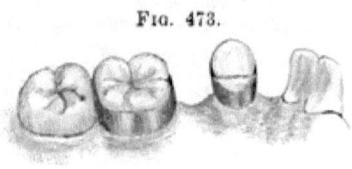

Fig. 473.

---

[1] *Dental Cosmos*, vol. xxviii, No. 12, page 745.

tenth of an inch in width, the collar is again driven on, and will appear as seen in Fig. 473. A stump corundum-wheel is then used to grind a shoulder on the lingual surface of the tooth, grinding also the edges of the collar flush with the shoulder. The collar is again removed, and a piece of thin platinum plate, about No. 32, sufficient to cover the lingual surface of the tooth, is caught on the lingual edge of the collar by the least bit of solder, and all put in place on the cuspid (Fig. 475). The platinum should now be burnished on to the shoulder, and over the tooth and collar to the extent shown by the lines in Fig. 475.

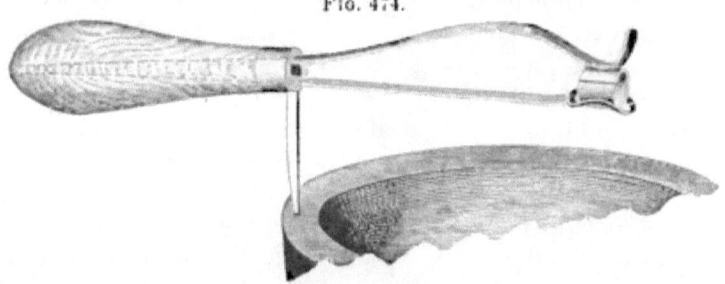

Fig. 474.

After trimming to those lines, and careful replacement and burnishing on the tooth, the collar and half-cap are removed, filled with wet plaster and marble-dust, and the platinum soldered to the gold. It is then placed on the tooth, burnished into all the inequalities of the tooth, very carefully removed, invested, and enough solder flowed over the platinum to cover and give it strength. Fig. 476 shows it complete on the cuspid.

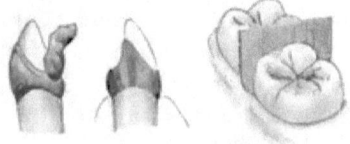

Fig. 475.   Fig. 476.   Fig. 477.

"I have feared that a detailed statement would imply a long and tedious process, but I have often made such collars in less than an hour, and in any case time must be made subservient to exactness of fit and adaptation to the end in view.

"In the preparation for fitting a collar on the first molar (Fig. 473), I should have wedged or otherwise separated it from the second molar, so that a piece of sheet brass might be put in place, as shown by Fig. 477, and an impression taken in plaster, which if allowed to get hard would bring away the metal. It not, it could be replaced in the plaster. Melted fusible metal, when near the cooling-point, is then poured into the impression, and when cold will allow the safe removal of both the plaster and the metal strip. On this metal model a collar can be formed that will accurately fit the molar, as seen in Fig. 473. If the molar has no antagonist, a cap may at once be struck up on the model; but if there be an antagonist the cusps of the natural molar should be removed by grinding at points where the occluding tooth will admit of sufficient thickness of the gold cap. An exact copy of the ground cusps can then be made in less than five minutes, by the use of moldine with its accessories, and the process is as follows: Make the tooth perfectly dry. Put the collar on it. Nearly fill the cup with moldine, and coat it with soapstone powder. Press the compound on the tooth and collar firmly to about one-fourth the depth of the tooth. Carefully remove the cup; trim off any overhanging material, and place the rubber ring over the cup to about one-half the depth of the ring. Melt the fusible metal and pour it as cool as it will run from the iron ladle. As soon as the metal is hard, remove it with the ring, taking care not to impair the impression, which can be used again if the die is found imperfect or gets injured in use. Place the die and ring in cold water, to remain until quite cooled. While the die is wet and held over a basin of water, pour into the ring fusible metal which has been stirred until it begins to granulate, and quickly immerse all in the water. The die and counter-die should separate readily by tapping them with a hammer, but if they stick others can be quickly made from the same impression by the same method, using more care. With this die and its counter-die, a piece of No. 29 or 30 gold plate is swaged to fit perfectly the cusps and collar, which, when removed, can be held to its place on the cap by the soldering-clamp, using spring pressure enough merely to hold them together for careful soldering with the pointed flame so as not to unsolder the collar. The seamless collars are excellent when care is used in selecting the proper size, as directed on the diagram (see page 246).

"The caps being in place on the cuspid and molar, an impression is taken with plaster; the caps accurately set in the impression, and hard wax melted with a hot spatula around the edges of the caps. The impression is then thoroughly coated with sandarac varnish, after which it is dipped for a moment in water, and filled with a wet mixture of one part marble-dust with two parts of plaster; using great care to perfectly fill the caps and molds of the teeth. Wait until this mixture has become quite hard; remove the cup, and with a suitable knife chip off the plaster without marring the cast; secure a good articulating

Fig. 478.

impression and transfer it to the cast to obtain an exact reproduction of the relative occlusions of all the teeth involved. With such an articulation in hand, and with the means already described for swaging gold or platinum plate to fit the cusps and articulating surfaces of either the natural or artificial teeth, it should be within the capacity of any competent dentist to complete a suitable bridge; although there are practical points that can only be imparted by clinical instruction and actual demonstration in the mouth. Such a bridge is shown in position by Fig. 478."

CHAPTER XIII.

PARTIAL CAP AND PIN-BRIDGE METHODS.

Dr. W. F. Litch's processes in this style of work have for special cases many decided advantages, and are thus described by him:[1]

"Fig. 479 represents a typical case, in which a lateral incisor (crown and root) has been lost, the cuspid and front incisor, fully vitalized, and without approximal carious cavities, remaining in position.

"*To Make a Pin and Plate Bridge.*—1. Take in plaster an accurate impression of the cuspid and incisor and the interspace. From this obtain a plaster model of the parts.

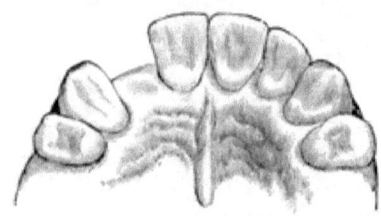

Fig. 479.

"2. Make from pure gold, rolled to the thinness of No. 26, standard gauge, base-plates, to be carefully adjusted to the palato-approximal surfaces of the cuspid and incisor. These can be made by swaging on dies and counter-dies obtained from the model, but more conveniently by bending the gold into shape upon the plaster model and pressing and burnishing it into perfect adaptation upon the natural teeth.

"3. Select a plain plate porcelain tooth of suitable length, shape, and shade, and wide enough to fit easily into the interspace. Let the neck of the tooth rest lightly upon the gum.

---

[1] *Dental Cosmos*, vol. xxviii, No. 3.

"4. With pure gold or platinum make a backing for the porcelain tooth.

"5. Place the tooth thus prepared and the base-plates already made upon the cast and accurately adjust the approximal edges of the base-plates to the backing of the porcelain tooth *in situ* upon the cast.

"6. When this adjustment is made, cement together the base-plates and backing with a brittle, resinous cement (resin, two parts; wax, one part; or sealing-wax will answer), and before the cement has fully hardened remove from the cast to position in the mouth, perfecting the final adjustment there. By this method much greater accuracy of adaptation is obtained, as the lines of length, width, and contour are too fine to be reproduced with absolute fidelity in a plaster model. In this part of the process too much care cannot be taken to have each piece of the appliance fitted with absolute accuracy to the surface for which it is designed. When this has been accomplished, throw upon the yet more or less plastic cement a stream of ice-cold water from an office syringe; this renders the cement perfectly brittle and incapable of bending. This done, immediately remove from the mouth and invest in a mixture of equal parts of marble-dust and plaster of Paris.

"7. After the investment has firmly set, solder the base-plates to the backing, and the backing to the platinum pins of the porcelain tooth, using as a solder 20-carat gold. Thus joined, the appliance will present the appearance shown in Fig. 483,—A representing the base-plate for the cuspid; B, the base-plate for the incisor; C, the porcelain tooth with its platinum backing; D, the points of union between the base-plates and backing. At these points the greatest strength is required, and it is important that here a large amount of the solder should be placed. The porcelain tooth being usually thinner than the natural teeth, there is nearly always an angle or depression at the points indicated, in which the thickness of the gold can be considerably increased without interfering with occlusion.

"8. For the purpose of attaching the denture as thus far constructed, drill a small cylindrical opening through the palatal surface of the enamel of the cuspid and incisor respectively.

These openings should usually be placed about as indicated in Fig. 482, at C, D. Sometimes, owing to a close occlusion or to the contour of the tooth, it is desirable that they should be located a trifle nearer the neck of the tooth. Each opening should be well undercut, but must not encroach upon the dentine far enough to endanger the pulp. In size the openings need not be larger than will admit a platinum pin-head, in diameter corresponding to No. 13, standard gauge, with a shank of No. 18, standard gauge. Into each of these openings must be fitted a platinum pin of the size indicated. The head of each pin must be made thin and perfectly flat both upon its upper and under surfaces.

"9. In each of the base-plates make an opening corresponding in position to those in the natural teeth. Pass through these openings and cement in them the free ends of the platinum pins.

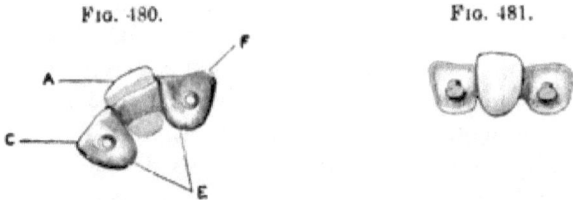

Fig. 480.   Fig. 481.

While the cement is yet plastic, place the denture in position in the mouth, carefully pressing the pin-heads into the openings made for them, and burnishing the base-plates into perfect contact with the palatal surfaces of the teeth; chill the cement, remove and invest as before, and with 20-carat gold solder the pins to the base-plates, flowing upon them and the backing as much of the solder as may be necessary to give them the desired thickness and rigidity; the amount admissible largely depending upon the nature of the occlusion; a central thickness of about No. 21, standard gauge, being all that is really requisite for strength, while the edges can be made much thinner.

"Fig. 480 represents the appliance without the pin. A is the porcelain tooth and backing; E, the base-plates; C and F, the openings for the pins.

"Fig. 481 represents the appliance completed with the pins in position.

"Fig. 482 represents the natural teeth and interspace B, with openings for retaining-pins, C, D.

"Fig. 483, already described, represents the appearance presented when the bridge is cemented in position.

"*To Attach the Bridge.*—To attach the bridge the best attainable oxyphosphate cement should be used. It is desirable that it should set slowly. Thoroughly dry the teeth and denture; mix

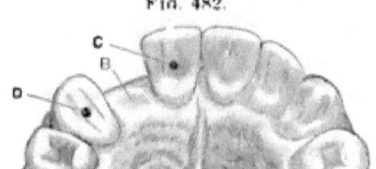

Fig. 482.

the cement to as thick a consistence as is compatible with perfect plasticity. A thick, viscid, semi-fluid mass is what is required. With suitable instruments, swiftly but carefully place the cement around the head and shank of each platinum pin, and also in the openings in the natural teeth. This care is necessary in order to exclude all air-bubbles and thoroughly engage the pin-heads in the cement. They furnish ample

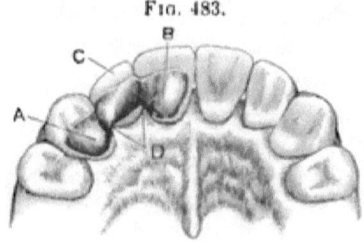

Fig. 483.

retaining surface, but none to spare. In packing the cement around the pins the under surface of the base-plates should at the same time be covered.

"The above details being perfected, the denture is at once carried to position, and with broad-pointed, serrated instruments pressed firmly into place, the excess of cement, if of the proper consistence, freely oozing at all margins."

"*Application to Pulpless Teeth.*—In the above description the vitality of the pulps of the cuspid and incisor has been assumed; but, as can readily be understood, the pin and plate bridge can be even more easily and securely placed when one or both pulps are devitalized, for the reason that, the pulp-chamber being empty, the pin-holes in that tooth can be made as much larger and deeper as may be deemed desirable, the size of the pin being, of course, correspondingly increased. In a devitalized tooth, too, the base-plates can be sunk into the palatal surface when they interfere with occlusions, as sometimes happens when the antagonism of the lower teeth is very close and the overlap is considerable.

"Ordinarily, however, such interference is inconsiderable, and the difficulty can always be overcome either in devitalized teeth by the expedient just suggested, or by carrying the base-plates as far away from the cutting-edge as practicable, at the same time making them at the point of contact as thin as is consistent with strength; finally, if necessary, removing a slight portion of the cutting-edge of the occluding lower tooth. . . .

"As a rule the writer has confined the use of this form of bridge to cases in which only a single incisor is missing, but he has successfully attached a front and lateral incisor to a cuspid and the remaining front incisor. Where an unusual strain is to be expected the retaining-pins and pin-holes should when practicable be made correspondingly large, or two smaller pins may be anchored in one tooth, which latter plan gives very great resisting power.

"*Pin and Plate Attachments to Bicuspids.*—Although chiefly applicable to the incisors, the pin and plate attachment may be successfully combined with crown or bar bridges for molars and bicuspids.

"Fig. 484 represents a practical case in which the upper third molar and the first bicuspid (both without antagonizing teeth) were utilized for the attachment of a bridge made of gold crowns with porcelain facings, to supply the loss of the intervening teeth.

"Fig. 485 represents the case as prepared for the bridge. A, the inner cusp of the bicuspid cut down to allow the placing of a sufficiently thick crown-plate; B, a cylindrical undercut opening between the cusps for a retaining-pin; C, the third molar, made uniform in size from neck to grinding-surface, the latter

Fig. 484.

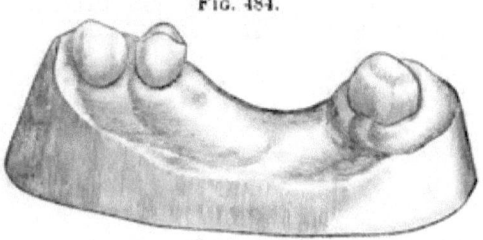

also being considerably retrenched; D, the crown-plate of a partial cap, made of pure gold, soldered with 20-carat gold, and so constructed as to cover every portion of the tooth except its

Fig. 485.

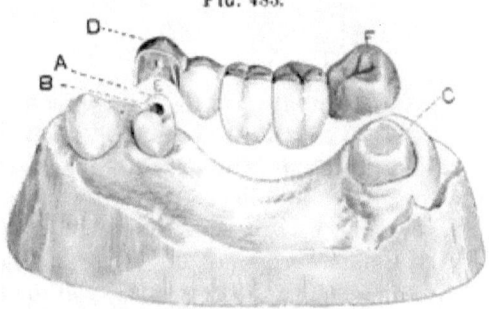

buccal surface, the free edge passing up under the gum; E, a retaining-pin adapted to the opening B; F, the gold cap for the molar.

Fig. 486.

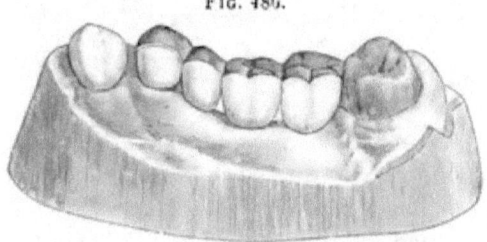

"Fig. 486 represents the bridge anchored in position with oxyphosphate cement.

"In the above case it will be observed that there is a considerable space between the bicuspid and cuspid. This made it readily practicable to give so considerable a thickness to the mesial wall of the partial cap as to hold it securely against the side of the tooth. Had the space been less, contact with the cuspid would have afforded the desired security.

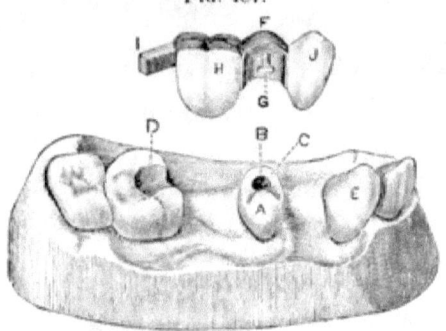

Fig. 487.

"Fig. 487 represents another case in which a bridge was attached by a bar, partial cap, and retaining-pin. A is an upper second bicuspid (without antagonist); B, its inner cusp, cut down; C, opening for retaining-pin; D, second molar, with slot for bar; E, cuspid; F represents the partial facing; G, the retaining-pin; H, a molar crown of gold, with porcelain front; I, a platinum bar attached to the crown (H) and made to fit into

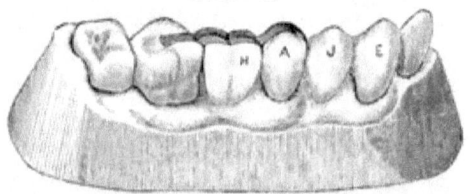

Fig. 488.

a slot (at D); J, a plain plate cuspid, heavily backed and strongly soldered to the partial cap, but left without attachment to or contact with the cuspid.

"Fig. 488 shows the bridge anchored in position.

"This case, after two years of wear, is still in perfect condition and doing good service. As it was possible to keep the gold attachments, backings, etc., out of sight, the appearance presented is very natural.

"The absence of antagonizing teeth for the bicuspids in each of these cases was a favorable condition, as a considerable thickness could be given to the crown-plate without any interference with occlusion. When the conditions are not so favorable, cutting down the inner cusp to the required extent and sinking the opening for the retaining-pin to the necessary depth are processes certainly to be, as a rule, preferred to the entire removal of the crown for the purpose of ferruling the root for the mounting of a crown of gold and porcelain,—a procedure, however, not by any means to be indiscriminately denounced, for in many cases it is in the highest degree advisable.

"There is this fact to be considered in regard to the use of the partial caps here figured,—that many patients can be induced to consent to their employment who would refuse to submit to more radical measures, and thus, even when the latter would be advisable, the former may be employed as a compromise, or even as a temporary expedient. Having once tested the advantage of a well-fitting bridge, the wearer is much more likely to consent to whatever measures are necessary to give it security and permanence."

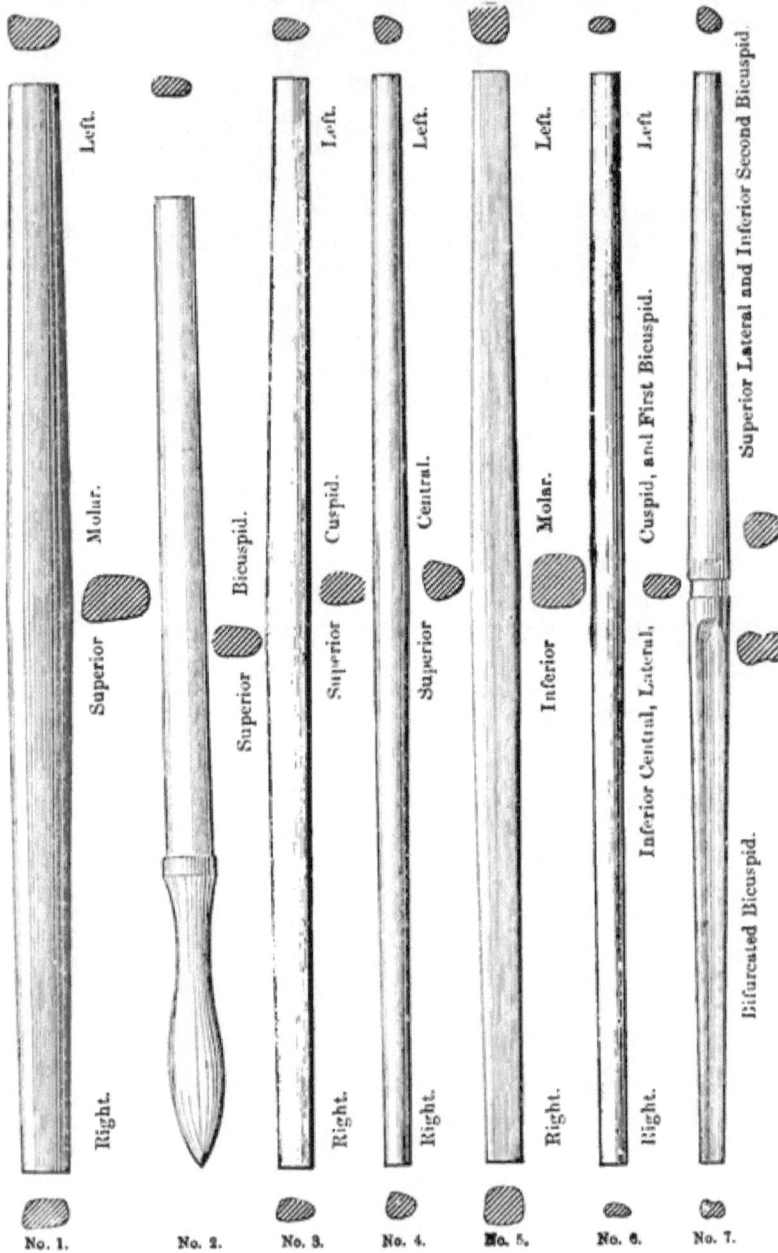

Fig. 489. MANDRELS FOR SHAPING SEAMLESS TOOTH-ROOT COLLARS.

CHAPTER XIV.

THE MANDREL SYSTEM.

In all styles of collar crowns the shaping and adapting of the collar is usually found by many dentists to be the most difficult part in their construction. To facilitate its performance the "Mandrel System" was introduced by The S. S. White Dental Manufacturing Co. The following is a revised description taken from an article on the subject:[1]

"An examination of a large number of human teeth shows that, no matter how great differences may exist in the apparent shapes of the crowns of individual teeth of a given class, there is a remarkable uniformity in the configuration of their necks. That is, the necks of upper cuspids, for instance, were found to have a fixed type, from which the variations were very slight as to shape, though there appeared to be no exact standard of size. So of the other classes, with the single exception of the superior molars, in which two distinct forms were found, the first being those in which the buccal roots were wider than the palatal; the second, those in which the reverse condition was found, the single palatal root being wider at its junction with the crown than the two buccal roots. The occurrence of roots of the second class being rather exceptional, the first class was accepted as the type.

"The configuration of the necks of all the teeth having been determined, a set of mandrels for shaping collars to fit them was devised. The set (Fig. 489) consists of seven mandrels, six of which are double end. Their shapes are modeled upon the general typal forms of the necks of the teeth which they represent, and they are made tapering to provide for all required variations in size. The illustrations are about two-thirds actual

---

[1] *Dental Cosmos*, vol. xxviii, No. 8.

size, the longest instruments being nine inches in length. The cross-sections show the shapes and proportionate sizes at the greatest and least diameters. The long taper permits the most minutely accurate adjustment of the collars.

"No. 1 is a double-end mandrel, for superior molars, right and left; No. 2 is a single mandrel, for superior bicuspids, right and left; No. 3 is double-end, for superior cuspids, right and left; No. 4, double-end, for superior centrals, right and left; No. 5, double-end, for inferior molars, right and left; No. 6, double-end, for the inferior centrals, laterals, cuspids, and first bicuspids, right and left; No. 7, double end, one end for the superior lateral incisors, the other for those bicuspids in which a bifurcation of the roots, or a tendency in that direction, extends across the neck to the crown in the form of a depression on one or both approximal surfaces. The foregoing scheme comprehends all the teeth of the permanent set except the second inferior bicuspids. The necks of these approximate those of the superior central incisors so closely in shape that it was deemed inexpedient to make a separate mandrel, as the No. 4 mandrel will serve for both.

"The collars or bands are made seamless, of No. 30 (American gauge) gold plate, 22 carats fine. Fifteen sizes, each of three widths ($\frac{1}{16}$, $\frac{2}{16}$, and $\frac{3}{16}$ inch) are made (Fig. 490), which it is believed will cover all requirements. These collars, although devised as a part of the system, can be used in all methods of crown- and bridge-work which require bands, and possess many advantages over any others. They are really labor-saving devices, as their use saves the time and trouble of making, and there is no danger of their coming unsoldered when the pins or the backing of the crown are being soldered; and there are no hard spots to give trouble in burnishing, as, for instance, close to the root, after the collar has been shaped and placed in position, the whole surface being uniformly soft.

"The seamless collars are also especially adapted to removable or detachable bridge-work. They are so constructed that Nos. 1, 16, and 31 exactly fit into or telescope with Nos. 2, 17, and 32, and so on through the entire set, each collar fits into the series next higher; so that a root may be banded with one size and

THE MANDREL SYSTEM

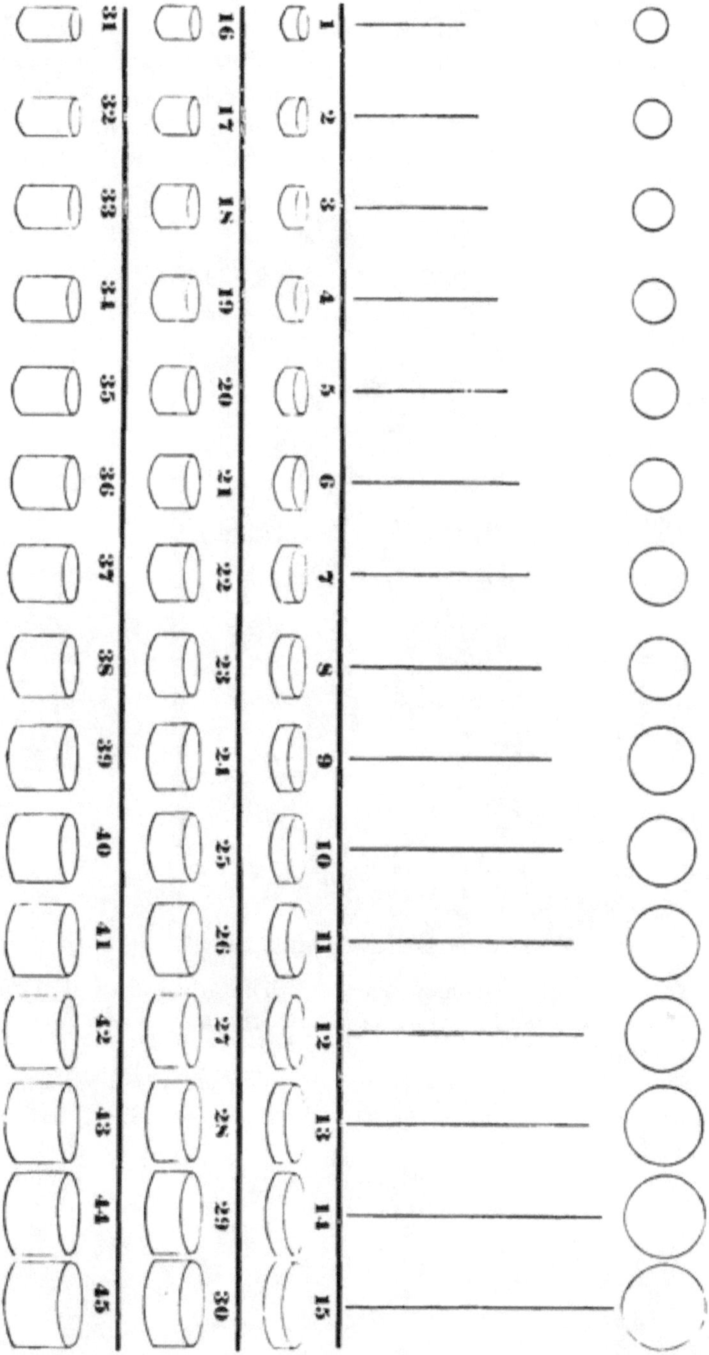

Fig. 450.

the size next larger used to form the tube for the telescoping crown. When desirable, the 'seamless' collar can be strengthened, after it has been adapted to the conformation of the crown so as to slide freely over it, by investing and flowing solder over the outer surface; or, still better, by adapting the next larger size of collar to exactly fit around the first, and then investing

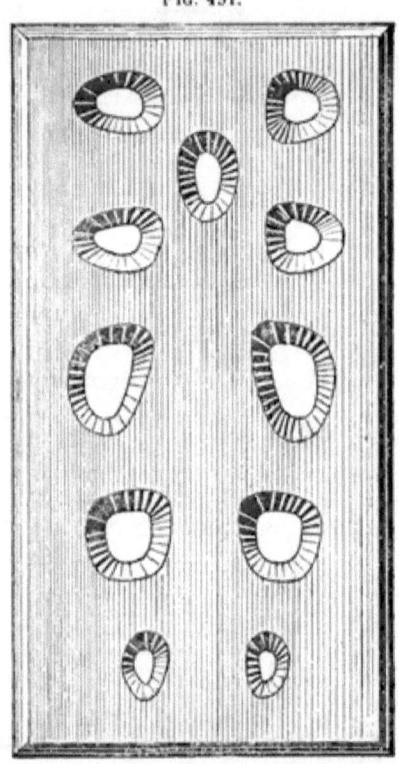

the two and soldering them together. The advantages of these collars for this kind of work, and also for the construction of cap crowns, are obvious.

"The other appliances specially devised for this system are, a reducing-plate or contractor, a pair of collar pliers, and a hammer.

"The contractor (Fig. 491) contains holes which are complementary in shape to the mandrels. The mandrels being applied to the inner circumferences of the collars, while the contractor must admit the collars themselves, the short taper of the holes in the contractor necessarily covers a somewhat greater range of

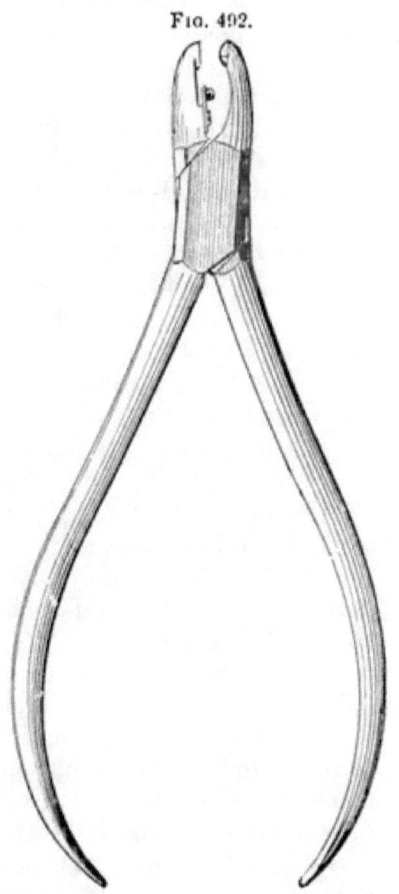

Fig. 492.

size than is shown in the mandrels. With this appliance collars can be evenly and accurately reduced in size at the edges, without burring or buckling. The illustration is actual size.

"The collar pliers (Fig. 492) are for contouring the collars to shape, one beak being made convex and the other concave to correspond. With this appliance the slightest changes required in the contour of the collars are easily made. About a half-inch from the extremity of the concave beak a small bar of flat steel is attached to it by means of a screw. The free end of the bar has a minute projection upon one face, the other being reinforced to fit into the concavity of the beak. In the center of the face of the convex beak is a depression, into which the projection on the steel bar strikes, making a very efficient punch for forming guards or stops to prevent the collars from being forced too far under the gum. The depression in the convex beak being slightly larger than the projection or punch, the metal is not cut through, but merely raised on the side opposite to the punch. The punch attachment being pivoted can be swung to one side when not in use.

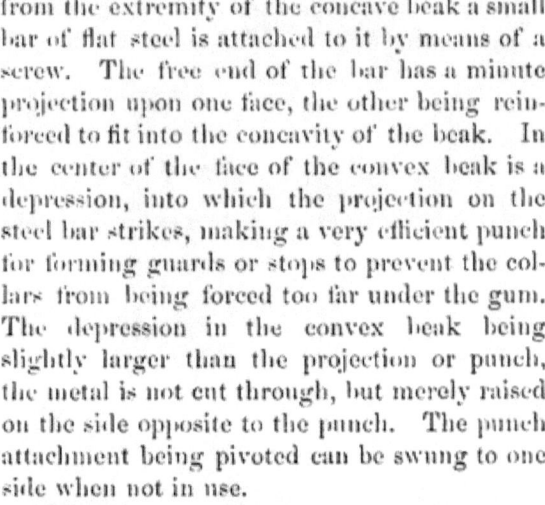

Fig. 493.

"Fig. 493 is a mallet or hammer, with steel face and horn peen. The handle is nine inches long.

"One of the appliances required is a lead anvil, which being only a piece of soft lead say two by three inches and an inch thick is not illustrated. The female die of an ordinary case will answer very well.

"To illustrate the uses of these appliances, take a case in which the two inferior bicuspids of the left side are missing, and the crowns of the cuspid and first molar so badly decayed that the probabilities are that they will soon fall victims to the forceps. The old-time way would have been to extract the molar and cuspid, and make a partial plate. Examination, however, shows that the roots of these two teeth are in good condition, affording an excellent opportunity for the construction of a piece of bridge-work.

"With a corundum-point or rotary file, cut off the remaining portions of the crowns level with the gum margins. Prepare the roots in any of the well-known ways, thoroughly cleansing the apical portions and filling them with whatever material is desired, being careful only that the work is well done. For the better retention of the filling-material to be placed in the pulp-chamber, retaining-grooves can be made or retaining-posts inserted. Take a piece of binding-wire (No. 26, American gauge), two and one-half inches long, pass it around the neck of the molar stump, cross the free ends, and, holding the wire in place with one finger, twist the ends with a pair of flat-nose pliers until the wire clasps the neck closely at every point (Fig. 494). Where there are any irregularities in the contour of the tooth, it is necessary to press the wire into them with an approximal burnisher. It is obvious that the ring thus formed will show the exact size and shape of the neck of the tooth. Remove the ring carefully, lay

Fig. 494.

Fig. 495.

it on the lead anvil, put over it a piece of flat metal, and with a smart blow from a hammer drive the wire into the lead (Fig. 495). Upon removing the wire an exact impression of the ring will be left in the lead anvil. (This part of the work, as indeed all others, should be done carefully as described. The wire ring may be driven into the lead by a direct blow of the hammer face, but the blow might not strike equally, and the interposition of the flat metal held level insures an even impression. A piece of an old file is best, as the file-cuts keep the wire from slipping.)

"Next, cut the wire ring at the lap, straighten out the wire, and select a suitable collar by comparing the length of the wire with the straight lines in the diagram (Fig. 496) which show the inside diameters of the various sizes. Should none of these correspond exactly, take preferably the next size smaller. It will be remembered that the collars are No. 30 in thickness, while the wire with which the conformation is secured is No. 26. This difference permits the collar when contoured to shape to enter the lead impression readily, a decided advantage in fitting. Having selected the collar, fit it to mandrel No. 5, with the peen of the hammer, holding it upon the lead anvil, and using a slight pushing force to help in stretching and forming it (Fig. 496). Having driven the collar to form, remove it from the mandrel and try in the lead impression. If it does not fit exactly, return it to the mandrel and stretch it a little, when it will usually fit perfectly, as the mandrels have been designed carefully to the average shapes which obtain in the great majority of tooth-necks. In the exceptional cases where the collar does not fit it can be readily contoured to the exact shape with a pair of flat-nose pliers. Of course, if it fits the impression in the lead, it will fit the neck of the tooth, always provided the measurement and the impression have been carefully made.

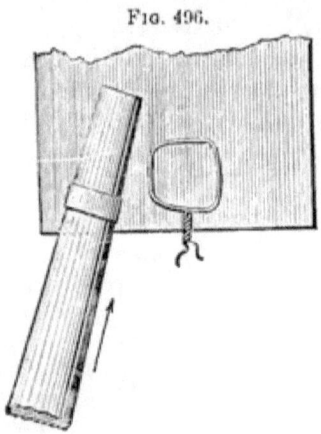

Fig. 496.

"If the collar or band has been accidentally stretched too much, or if for any reason when brought to shape it is too large, its root end can easily be reduced to the proper size by

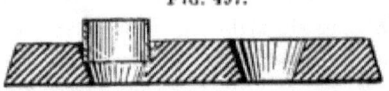

Fig. 497.

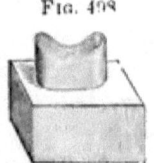

Fig. 498.

the use of the contractor. Place the edge of the collar which is to fit the root in the proper hole; hold it level with a piece of file as in taking the lead impression of the ring, and tapping lightly on the file drive the collar into the plate (Fig. 497) until the proper reduction is made. The collar is next 'festooned' to correspond to the shape of the maxillary ridge. Lay it, gum edge up, on the lead anvil, and with the piece of flat file and the hammer drive it into the lead. A few cuts with a fine half-round file across the approximal diameter will conform the edges to the surface of the ridge (Fig. 498). Then place the collar in position, and, having ascertained just how far it should go down on the root, remove it, and with the small spring punch in the collar pliers form projections on the inside of the band at the proper points to serve as stops, which, resting on the top of the root, will prevent the collar from being forced further down upon it than is desirable (Fig. 499).

"A collar for the cuspid is then fitted in the same manner, using mandrel No. 6 for shaping, after which the case is ready for the building of the bridge.

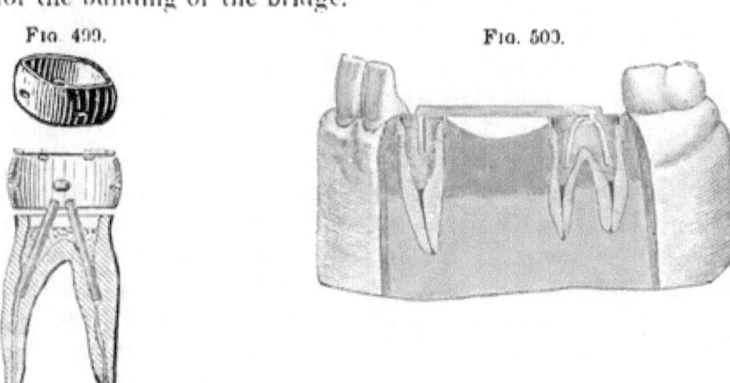

Fig. 499.   Fig. 500.

"Cement with oxyphosphate both collars in position. Then take an impression of the parts, including the interiors of the excavated pulp-chambers, from which make a cast in the usual way. Bend a short piece of half-round gold or platinum wire into the form of a horse-shoe, the two extremities of which shall fit into the roots of the molar. Then take a longer piece of the same wire, somewhat more than enough to extend from the toe of the horse-shoe when in position to the cuspid root; bend one end of it at a right angle, or nearly so, to fit the root of the cuspid, and (cutting off any excess of length) solder the other end to the toe of the horse-shoe. The bar extending between the two roots is the truss of the bridge. Next, place the appliance on the cast (Fig. 500), holding it in position with wax, and select the teeth to take the place of the missing bicuspids and molar. The best form for this purpose is a tooth having holes extending through it vertically from the neck to the grinding-surface similar to the well-known Bonwill crown.

"The crowns used should be large enough to fill the space rather tightly, even if their sides have to be flattened slightly to let them in. If the teeth do not fill the space tightly, a small portion of plastic filling-material crowded between them, as mortar between the granite blocks in the arch of a railway bridge, will greatly increase the strength of the work.

"After the teeth are ground to fit and the proper length for occlusion is ascertained, the truss is covered with a thin film of wax, upon which the crowns are again pressed to their positions. Upon the removal of the crowns the impression of the holes running through them will be found in the wax. At these points drill holes through the bar with a small twist-drill run by the engine, and into these fit and solder the pins for the support of the crowns.

"The bridge is now ready to be attached permanently. Set the crowns in position upon their supporting pins to secure the proper alignment. (If the operation were upon the upper jaw they would have to be held with wax.) Put into the canals of the supporting roots (the cuspid and first molar) a sufficient quantity of some quick-setting plastic, as oxyphosphate, to about half fill the pulp-chamber, but not enough to prevent the supports of the truss from being forced home. Force the bridge supports to place, and after allowing the filling-material to become set remove the crowns. Fill the remainder of the pulp-chamber and the whole of the collar with gold or with amalgam, gutta-percha, oxyphosphate, or any suitable plastic (Fig. 501). Set the crowns permanently, the molar and cuspid first, as this affords greater facility for the trimming off of any excess of the filling-material used in the attachment. For attachment of the crowns, gutta-percha is probably the best material, as crowns set with it are readily removed for the correction of any inaccuracies of occlusion or alignment, by grasping them between the beaks, previously warmed, of a pair of universal lower molar forceps. The heat warms the gutta-percha and releases the tooth, which can then be reset properly. In attaching crowns with gutta-percha the holes in the crowns are first filled with the material, after which the crown is warmed and forced to place. Any of the other plastics ordinarily used in setting

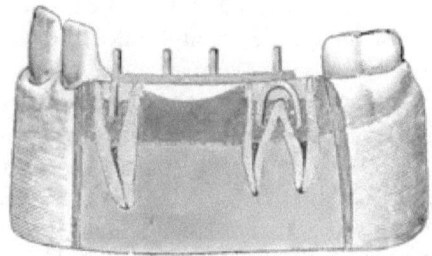

Fig. 501.

Bonwill crowns can be employed, at the discretion of the operator. Fig. 502 shows the case completed.

"In securing the occlusion of a piece of bridge-work it is well to make the artificial teeth a little short, so that the natural teeth on both sides will meet the first shock of mastication. Nature will correct the occlusion in time by slightly elongating the roots supporting the bridge. If the artificial crowns are permitted to strike the natural teeth from the first, the undue strain

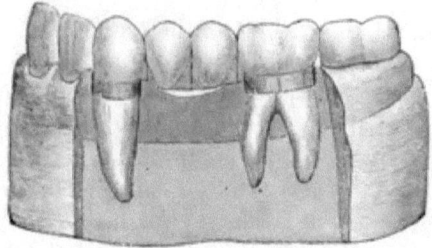

Fig. 502.

upon the two supporting roots may cause soreness and perhaps more serious consequences.

"When a sound tooth is to be used as one of the supports of the bridge, a modification of the method just described is necessary. Take a case where it is desired to bridge the space caused by the loss of the right inferior bicuspids and first molar. The crown of the right cuspid is nearly gone, but the root is sound and capable of supporting one end of the bridge. The other end will be attached to the second molar, which is a sound tooth. Prepare and band the cuspid root as before; dress off the second molar crown until it is slightly smaller than the neck, and shorter at the occluding surface, so as to permit a cap to be telescoped over it, and take the measure of the crown with the binding-wire. Select a suitable seamless collar of sufficient width to extend from the neck to a little beyond the grinding-surface, and drive it up on the proper mandrel to get the general shape, but not the full size required to fit the tooth, leaving it so that the edge having the larger circumference will just pass over the end of the crown; place the collar on the tooth, and with a block of wood and the mallet tap it to place

FIG. 503.

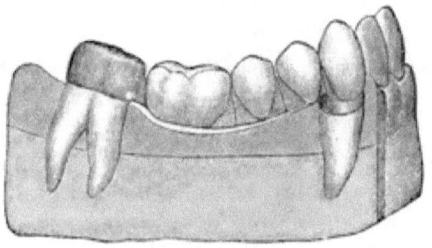

just beyond the free margin of the gum. This method will make a close fit, as the collar will readily stretch all that is necessary. With a sharp-pointed instrument mark the length of the crown and line of the gum margin, remove the collar, and cut it to the proper form as indicated. Then in a piece of gold plate of the thickness used for caps form four little depressions of the general character of an impression of the molar cusps. An easy way to do this is to lay the plate on the lead anvil; then with the ball on the end of an ordinary socket-handle and the hammer the depressions are made in a moment. Clamp the collar on the plate, borax it, charge with solder, and heat till the solder flows. Cut off the surplus plate, and a perfect cap for the molar is made. Place it on the tooth and take an impression, and thereafter proceed as before directed to make the truss of the bridge and mount the teeth, except that in this case the posterior end of the truss is to be soldered to the molar cap. For the final attachment place a little oxyphosphate in the cap to secure it firmly (Fig. 503), first cutting a slot in the crown end of the cap for the escape of the excess of material. Pressure upon the filling-material hastens its hardening." . . .

DETACHABLE BRIDGE-WORK.

"A method of constructing a detachable bridge applicable to cases where one or both of the supports or piers are sound teeth is as follows: In the case adduced for illustration the right inferior cuspid crown was decayed, and both of the bicuspids and the first molar were absent. The supports for the bridge were the sound second molar and the cuspid root. After the cuspid root was prepared and banded, the crown of the

Fig. 504.

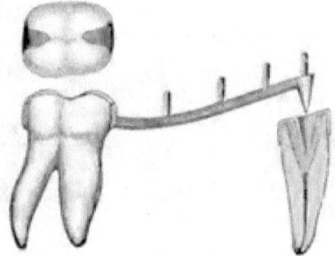

molar was reduced very slightly,—not sufficient to destroy the enamel, but just enough to permit a collar properly fitted to pass over it. A collar somewhat wider than the length of the crown from grinding-surface to neck was fitted and cut to the proper width. Two lugs were then soldered upon the anterior and posterior sides and bent to fit into the approximal fissures, which were slightly cut out to admit them. An impression was taken, the collar coming away in the plaster, and a cast was made with the collar in position. A coned tube was then made for the root of the cuspid and a coned pin fitted into it. A truss of half-round wire was made, to which the coned pin and the molar collar were soldered (Fig. 504). A half-clasp to grasp the lateral was next soldered to the end of the truss to be supported by the

17

cuspid. The object of this clasp was to guard against the teeth being thrown out of proper alignment by the force of mastication. Bonwill crowns were then vulcanized to the truss, after their supporting pins had been fitted and soldered to it. (Countersunk crowns can be used as well in the same way. Plain plate teeth may also be used in this style of work, in which event they are to be soldered to the truss.) The bridge was then ready to be set, which was accomplished in the following manner: The cuspid root was nearly filled with oxyphosphate, and the coned tube was placed upon the pin. The band was put on the molar, and the coned pin with the tube upon it was forced into the plastic in the cuspid. As soon as this became set, the tube was held permanently, while the bridge itself could be removed whenever desired (Fig. 505).

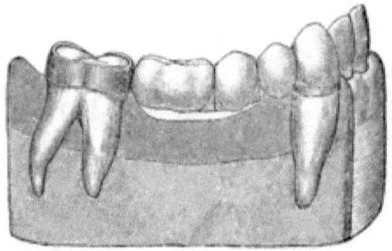

Fig. 505.

"This method of fixing the tube allows considerable range in its adjustment. In soldering the coned pin to the truss, care should be taken to set it at an angle exactly parallel to the axis of the molar; otherwise there will be difficulty in removing the bridge.

"A second style of detachable bridge-work to be described involves the use of cusp crowns (Fig. 506) for supporting posts or piers. Suppose a case where a bridge is required to extend from the right inferior cuspid to the right inferior second molar, with only the roots of the two teeth named as supports. Prepare the roots and pulp-chambers. Set screw-posts into the dentine for anchorage or as retaining-pins, and fit the collars, using sizes wide enough to form the walls of the crowns. Fill the pulp-chamber and cement on the collars, filling about two-thirds of the depth of the collars with a plastic filling-material, packing it well around the retaining-posts. Select suitable cusp crowns for the molar and cuspid, and place them in the ends of the bands to ascertain the occlusion. If too long, shorten the cusps or reduce the bands with engine corundums or rotary files, and when the correct articulation is found form a small square shoulder in the lingual edge of the cuspid and in the posterior grinding-surface of the molar. Fill the remaining portion of the collars with plastic mixed somewhat thinner than the first lot, and set the cusp crowns in position. If there are antagonizing teeth the mere closing of the patient's jaws will force the crowns to place. If there are no antagonizing teeth the crowns can be readily tapped to place with the mallet, using a piece of wood as a driver. Allow the filling-material to set firmly, trimming off any excess which may exude around the collars.

FIG. 506.

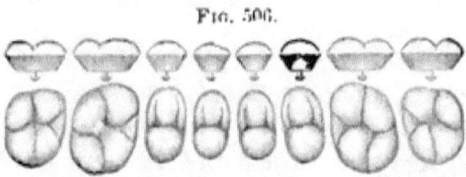

"Bridge supports or piers constructed on this plan are strong and durable, and likely to withstand any strain. Take an impression, and proceed to fit seamless collars to telescope over those already set upon the cuspid and second molar roots. It will be remembered that these collars are so made that each size telescopes into the next higher series. If the proper sizes are selected for the outside or female bands, the work of fitting is readily and quickly accomplished, forming tubes which slide easily over the supporting piers, and at the same time fit closely. It is only necessary to take care in shaping the tubes not to drive them too far up on the mandrels and thus stretch them so as to destroy the fit. To the outer end of each of the tubes solder a small piece of gold plate, forming partial caps so placed as to rest when in position upon the shoulders previously cut in the cusp crowns. Adjust a truss bar of half-round gold wire, to the ends of which solder the tubes. The truss is now ready for the teeth, which may be of any of the forms used for this purpose, and they may be attached to the bar in any way desired. One of the strongest attachments is vulcanite. Fig. 507 shows the construction and the finished case.

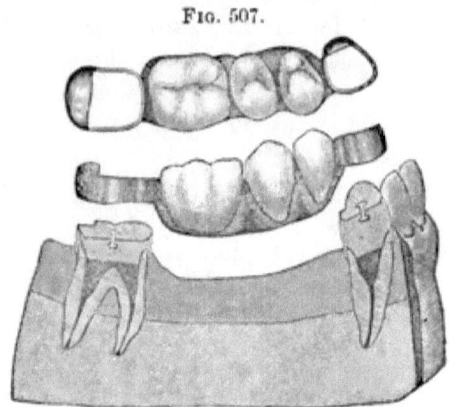

FIG. 507.

"An easy modification of the plan just described is readily adapted to cases where only a small space is to be filled and one end of the bridge is to be supported by a sound tooth. Thus, suppose it is desired to bridge a space formerly occupied by the two inferior left bicuspids, the crown of the first molar being a mere shell. The operation would be essentially the same as in

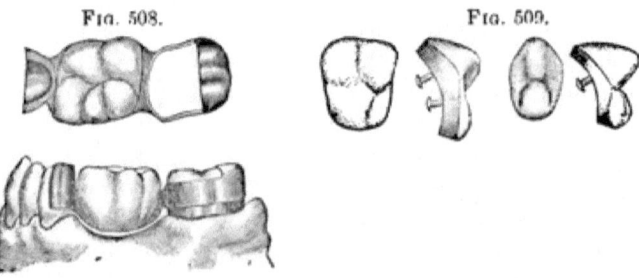

FIG. 508. FIG. 509.

the previous case, except that the sound cuspid would be utilized for one of the piers as follows: Fit a seamless collar, cut out a portion of it so that it will embrace only about two-thirds of the cuspid crown, and solder a partial cap or cover to it, as illustrated in Fig. 508. Or, if deemed preferable, the cuspid may be separated from the lateral incisor with the corundum-disk and the collar allowed to embrace the whole crown.

"A crown broken from a bridge constructed by any of the methods above described can be easily substituted, and the piece when repaired will be as strong and serviceable as it was originally.

Fig. 510.

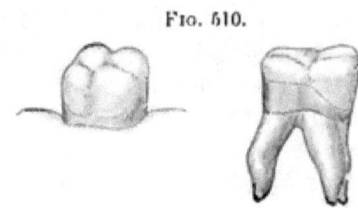

"It has not been deemed necessary to detail the construction of a single crown separately, as all the steps are included in the building of bridges, which have been described minutely. Porcelain cusps of the general form illustrated in Fig. 509 have been designed specially for these cases. In mounting them the gold band is cut away on the buccal side as shown in Fig. 510 to permit the porcelain to show."

CHAPTER XV.

PORCELAIN BRIDGE-WORK.

For an ideal piece of bridge-work that will dispense with the objectionable features of a permanently attached bridge and refute the most forcible arguments against the system, porcelain is the most suitable material known.

DR. BROWN'S METHOD.

Dr. E. Parmly Brown, of Flushing, N. Y., has invented and introduced a method of porcelain and iridio-platinum bridge-work, which possesses special merit.

The advantages claimed for it as a method of bridge-work are as follows:

It is formed entirely of porcelain with an iridio-platinum bar running through the denture as a sustaining shaft, thus presenting a perfect continuity of porcelain surface.

It is unaffected by any chemical action or condition in the mouth. It has no so-called self-cleansing spaces, because none are required.

A benign and natural form of contour is presented on its palatal surface, as the base of each tooth presses tightly onto the membranes of the gum, which closes or hugs closely around it.

And lastly, it has the merit of simplicity of construction in comparison with the other methods in practice.

Construction.—The bridge is formed by spanning intervening spaces between certain natural teeth or roots with artificial porcelain substitutes baked onto an iridio-platinum bar; either or both ends of the bar being anchored in a crown, or in a filling inserted in the approximal portion of the adjoining tooth. The roots to be crowned and used as foundations for the bridge are prepared as for single crowns. If the end of the bar is to be fixed in a natural tooth, the cavity that is to receive and anchor it is opened up sufficient to admit it,—in the bicuspids and molars from the grinding-surface to the cervico-approximal edge, and in the incisors, on the approximal side, with an opening on the palatal or labial face, as indicated. A square bar of iridio-platinum wire from No. 13 to No. 15, U. S. standard gauge, in thickness, is fitted either to the cavities of the teeth in which it is to be anchored, or bent and fitted in any devitalized tooth or root which is to act as an abutment. The end of the wire that forms the post is pointed and introduced well up the root-canal, and the end forming the bar is slightly flattened or squared. Any intervening root-posts are fitted and riveted to the bar as it passes above them.

To this bar the teeth used, which are ordinary plate teeth,[1] are fastened: incisors by slightly flattening the bar and riveting them fast, or when straight-pin teeth are used by bending the pins over the bar (Fig. 511); bicuspids and molars with straight pins, by grinding a slot with a disk between the pins and bending them over the bar, which is slightly barbed and set in the slot.

Fig. 511.

The proper position of each tooth having been determined by adjustment in the mouth, or to the articulating model, the rivets and bar on each tooth are carefully filled around and the cervico-palatal portion rounded off with porcelain body so as to present a natural surface to the tongue. The shape of the end of the root is given to the crown by placing the porcelain body, to which a small quantity of starch has been added, around the post as it is fixed in position on the model: the plaster having been previously varnished with collodion to prevent cohesion. The spaces between the teeth on the line of the bar are to be substantially joined with the porcelain body; free spaces between the teeth at the cervical portion are to be preserved as much as possible without interfering with strength or producing places that will be uncleanly.

---

[1] Dr. Brown uses The S. S. White Dental Mfg. Co.'s teeth entirely for this work, as he has found them the most suitable, being the only kind that will retain their color in the intense heat to which they are subjected.

The bridge is then placed on a slide in the muffle of a continuous-gum furnace. It is sustained in position by inserting the pins in holes drilled in the slide (Fig. 512) or by suspending it with platinum wire between two platinum posts inserted in the slab. The bridge is then baked the same as continuous-gum work. It is next fitted to the mouth. If any alterations are required they should be made, or if any imperfections in the body occur the places should be filled in and the bridge again baked. It is then ready for insertion.

Insertion.—When the bridge is supported by crowns alone, the posts are barbed and the bridge is then cemented on, the same as any other. If one end is supported by a porcelain crown and the other by a bar, the filling on the bar end is inserted up to the

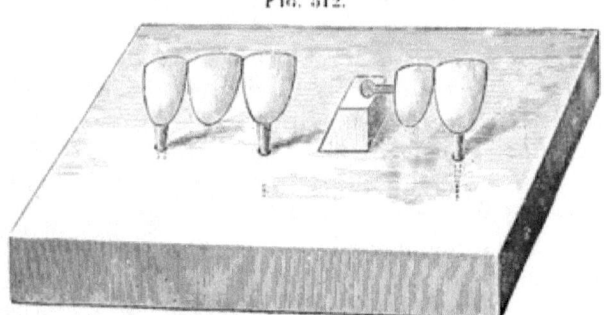

Fig. 512.

position the bar is to occupy before fastening the post of the crown end, which is then cemented and the filling of the bar end completed. If both ends are bars, both fillings are inserted up to the position of the bars, the bridge inserted, and the fillings finished; the bridge being held by an instrument or the fingers until the bar ends are partly covered.

When gold is used, the rubber-dam should first be applied, and the cervical portion of the teeth of the bridge pressed tightly into it to avoid any space being left after its removal and the completion of the operation.

When amalgam is used, the rubber-dam is not necessary.

In Fig. 513, No. 1 is a third molar, pulp alive, with large filling; No. 2 is a porcelain bridge; No. 3 is a first molar, pulp dead, with a metal bar entering the pulp-chamber.

PORCELAIN BRIDGE-WORK.   265

In Fig. 514, No. 1 is a second molar, pulp alive, with a crown filling of gold or amalgam retaining the bar; No. 2 is a porcelain bridge; No. 3 is a gold crown with bar passing through one side of the crown into the root.

Fig. 515 is a view of a bridge of two teeth,—a central porce-

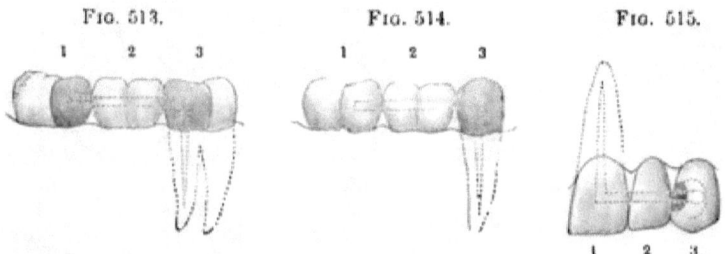

lain crown with a lateral baked onto it, the bar and pin being of the same piece, bent at about a right angle. No. 1 is a porcelain crown forming part of the bridge; No. 2 a bridged lateral with metal bar baked through it; No. 3 a living cuspid with a metal bar running into the center of a solid gold filling.

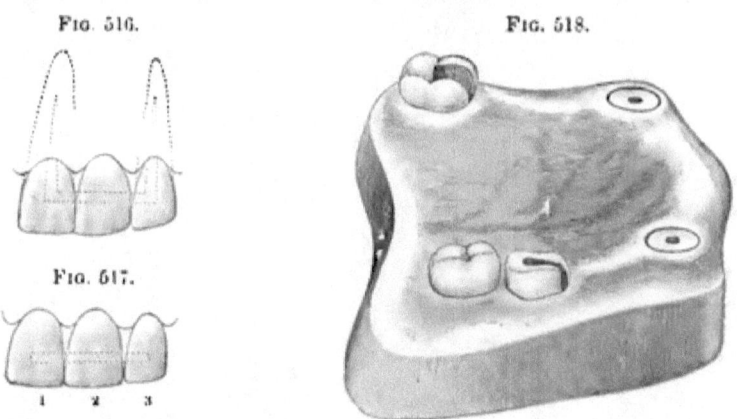

Fig. 516 is a view of a central incisor bridged onto two teeth whose pulps have been lost.

Fig. 517 is a view of an extension bridge consisting of a right central and left lateral, supported by a left central tooth or crown as the case may be. The bar can be anchored in a filling in the natural crown or attached to the artificial one. Nos. 1 and 3 are teeth on a porcelain bridge; No. 2 the natural tooth or artificial crown on which the bridge is saddled.

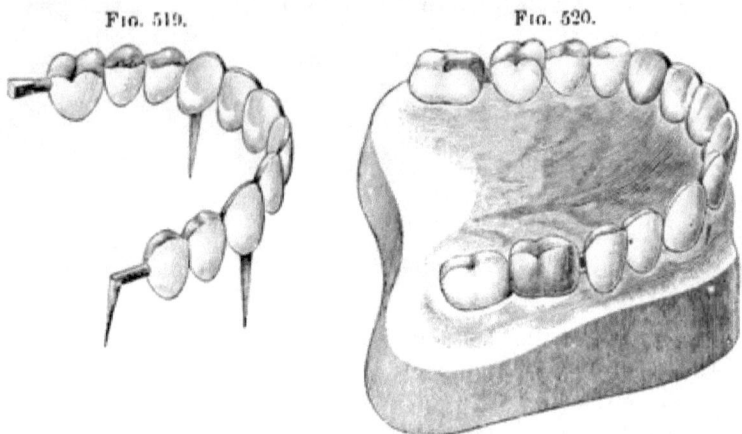

Fig. 519.   Fig. 520.

Figs. 518, 519, and 520 represent a bridge of eleven teeth recently inserted by Dr. Brown on two cuspid roots, a pulpless molar on the left side which was capped with a gold crown,

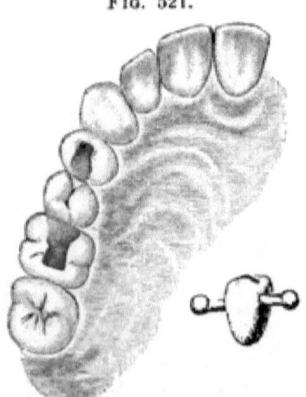

Fig. 521.

through the side of which the bar was passed and anchored in the root, and a molar on the right side into which the other end of the bar was anchored in a gold filling.

Fig. 521 represents a bicuspid bar-bridge anchored in the molar and first bicuspid. The side cut shows the construction. This operation was performed by Dr. Brown for Dr. Wm. Crenshaw, of Atlanta, Georgia, at the anniversary clinic of the First District Dental Society of the State of New York, in January, 1887.

Fig. 522 represents an extension bridge in which the abutment consists of a crown and bar combined.

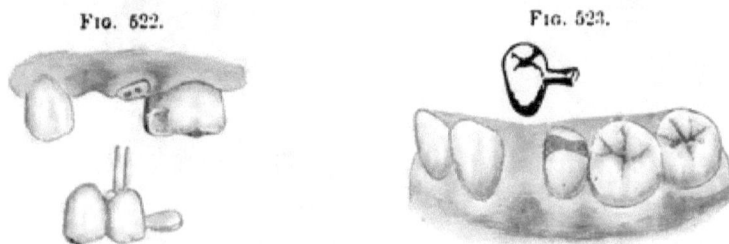

Fig. 522.   Fig. 523.

Fig. 523 represents an extension bar-bridge. The filling in the second bicuspid, which extended from the mesial to the distal side, had been inserted some time previously. Enough of the gold was removed from the mesial section of the filling to

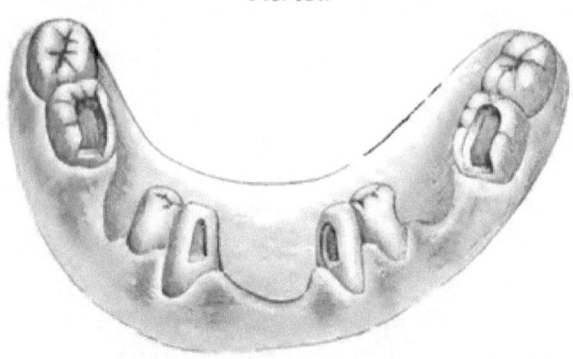

Fig. 524.

admit the bar, which was then securely anchored and the filling restored to its original form. The operation was performed by Dr. Brown for Dr. F. P. Hamlet, and to present date has been worn three years. The antagonizing teeth in the act of occlusion favor the artificial tooth forming the bridge.

In porcelain bridge-work, should a case require it, artificial gum in a moderate amount can be formed above the teeth of the bridge, to restore the contour of the parts. Figs. 524 and 525 illustrate a case of this character. A, Fig. 526, shows the

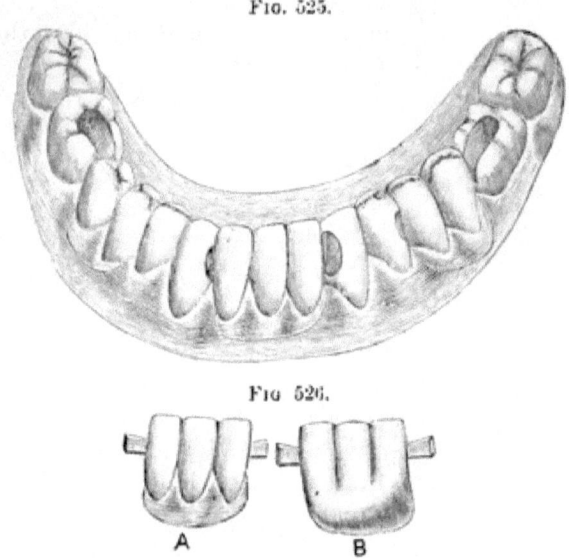

Fig. 525.

Fig. 526.

labial and B the lingual aspect of the incisor bridge before insertion.

In forming this style of porcelain bridge, when it is considered

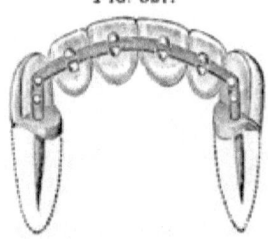

Fig. 527.

preferable to cap the roots, platinum plate can be struck up for the purpose as described at page 105, and after being adjusted to the ends of the roots the posts are passed through them, as illustrated in Fig. 527. Cap and bar are then soldered together with a very small quantity of pure gold, and the construction of the bridge continued.

Fig. 528 represents a case recently restored by Dr. Brown's method by Dr. M. L. Rhein, of New York. To remedy the abnormal character of the occlusion, the lower anterior teeth and the right upper cuspid were trimmed to the dotted line seen

FIG. 528.

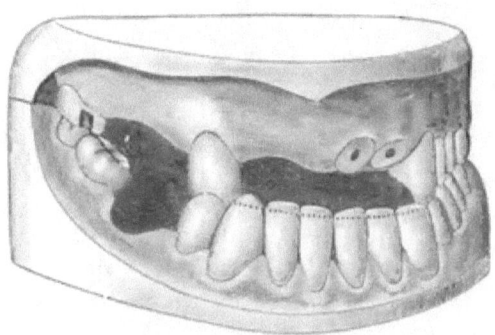

in the figure. To replace the missing teeth porcelain bridge-work was then constructed. The lateral and central roots, and the right cuspid and the molar, constituted the abutments. On the central and lateral roots were mounted caps with collars. A platinum crown was then made for the cuspid (Fig. 529), and to this crown was attached the bar, which was extended to its anchorages in the molar crown and the lateral and central roots,

FIG. 529.   FIG. 530.

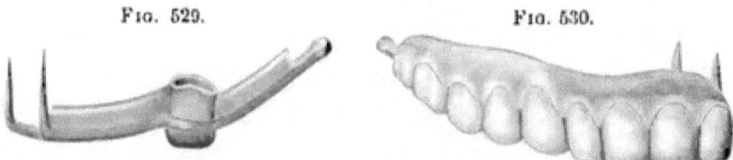

the caps on the ends of which the posts pierced. Owing to the large quantity of porcelain body to be used in forming the artificial gum, a strip of platinum plate was extended above the bar to stiffen its projecting ends and prevent warpage in baking. The artificial teeth were then articulated to meet the incisive edges of the inferior natural teeth, and thus in a measure overcome the deformity caused by the abnormal occlusion. In doing this the labial aspect of the cuspid was covered by the artificial teeth. Porcelain gum was then formed above the teeth in sufficient quantity to restore the contour of the parts. It was brought to a feather

FIG. 531.

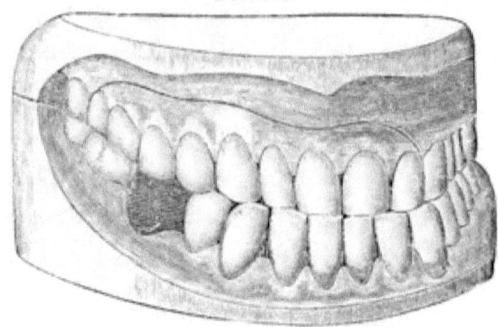

edge and pressed hard against all the tissues to insure a hygienic condition. Fig. 530 shows the finished bridge. In Fig. 531 the bridge is seen in position. The artistic result and improved

FIG. 532.

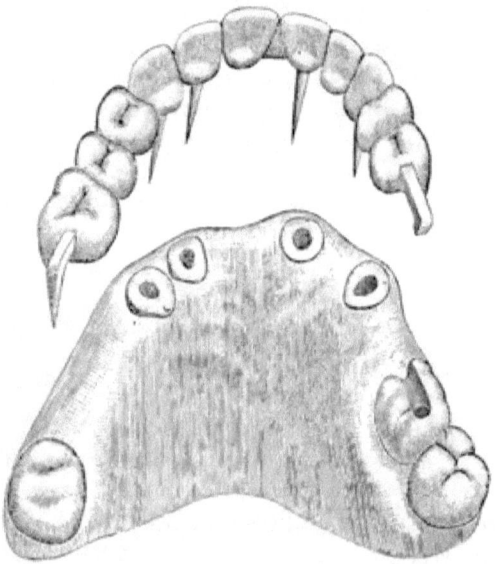

appearance effected are at once apparent. The mechanical construction of the piece was intrusted to Dr. C. L. Andrews.

Figs. 532 and 533 represent an upper and lower case of porcelain bridge-work, inserted by Dr. Wm. Crenshaw, in which several roots and teeth were used to form the necessary abutments. The upper section anteriorly receives the combined support of several roots as illustrated in Fig. 532. The bar on the left side is made flat and hooked on the end, affording a secure means of anchorage. A gold crown forms the support on the right. The lower section is firmly supported on the right

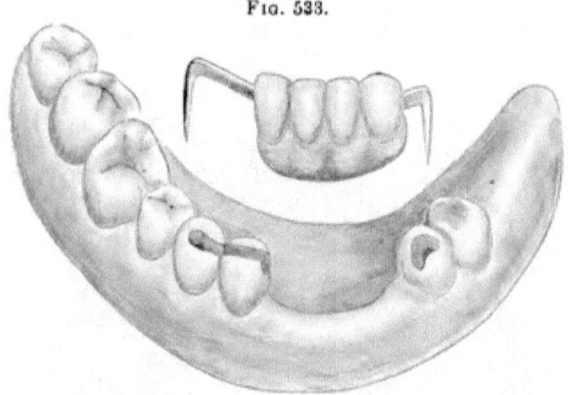

Fig. 533.

side by passing the bar through the cuspid and inserting the end in the pulpless bicuspid as illustrated in Fig. 533. The pulpless bicuspid on the left constitutes the other abutment.

Dr. Brown's system and method of bridge-work compares with other forms of bridge-work as continuous-gum work compares with other forms of plate-work. Its cleanliness, through the unalterable character and continuity of its surface and the incorruptibility of the material, renders it a superior form of denture. The ultimate value of this method, however, as a means of replacing lost members of the dental arch, depends, like others, on its proper, correct, and skillful application to cases suitable for it.

CHAPTER XVI.

CROWN- AND BRIDGE-WORK COMBINED WITH OPERATIVE DENTISTRY IN DENTAL PROSTHESIS.

The combination of operations on the natural teeth with crown- and bridge-work affords extraordinary advantages in dental prosthesis. The results which can be accomplished commend the plan strongly to the experienced practitioner. A few cases are adduced in illustration.

In the case presented in Fig. 534 the operative procedures were confined to the upper jaw, the lower teeth of the patient being

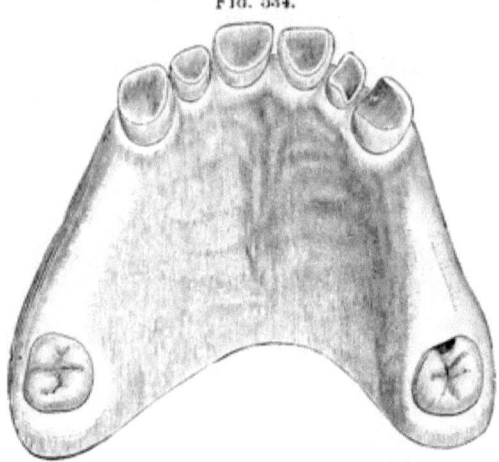

Fig. 534.

in good condition. The bicuspids and the first and second molars of both sides of the upper jaw had been lost many years before, and the incisors and cuspids showed the effects of extensive abrasion. The occlusion was sustained and the principal part of mastication performed by the incisors, as the third molars had been forced backward and antagonized only very slightly on one side.

The patient, a gentleman, had had a plate inserted, to the presence of which he had vainly endeavored to accustom his mouth. The abrasion of the incisors and cuspids was of the rapidly progressive character. These teeth were contoured with gold foil to

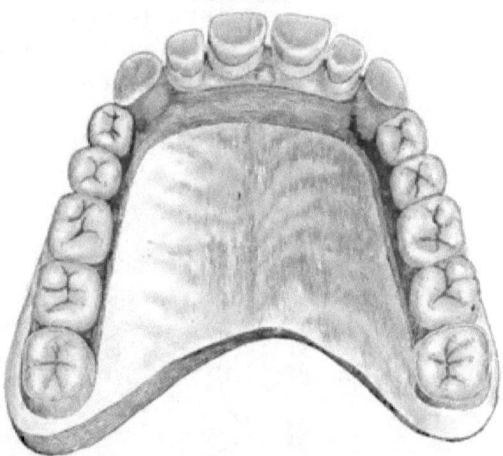

Fig. 535.

the form shown in Fig. 535, and a wire post was inserted in the right lateral, which was pulpless. At the occluding section of each filling, the layer of gold, after being packed with the plugger, was additionally condensed and hardened with a Herbst agate-

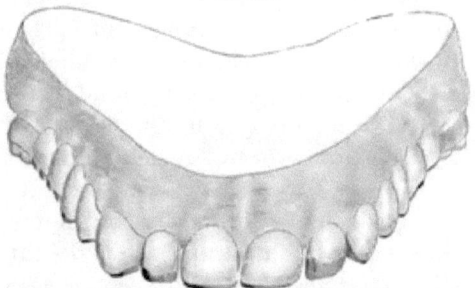

Fig. 536.

point burnisher to enable it to better resist the aggressive force of the lower teeth. The third molars were capped and lengthened with gold crowns, the lines of their sides being made parallel, to admit of a proper adjustment of the supporting collars for a removable plate bridge, by adding gold on their external surfaces. The plate bridge was employed because of the space between the abutments, which suggested the idea of utilizing the alveolus to assist in supporting it. A narrow shoulder was formed on each crown to support the collars. The attachments to the cuspids rested by means of a little shoulder on the occluding portions of the fillings inserted. Fig. 535 shows the completed denture, and Fig. 536 an anterior view of the same.

Fig. 537.

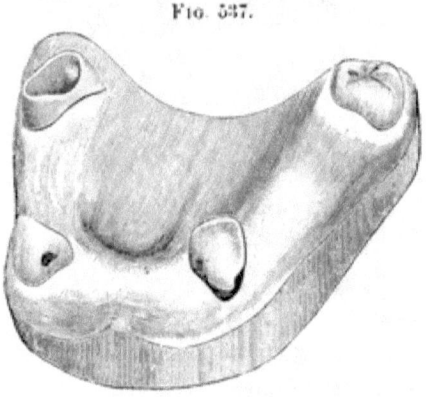

In the case illustrated in Fig. 537, the patient, a lady, had previously worn a plate, the clasp of which had entirely abraded the sides and cervix of the right cuspid of enamel, and caused recession of the margin of the investing gum. The decay which followed the abrasion extended in a circle around the tooth. As the patient objected to crowns of any kind being applied to either of the cuspids, the decay was removed and the edges of the cavity given a retaining form. A gold filling was then introduced in three sections, two of which embraced the approximal and palatal sides, while the third surrounded the labial wall, joining the other two sections at that point, the three thus completely encircling the tooth with gold. A portion of the filling was brought over the edges of the cavity to better shape the tooth for the attachment to be applied and also to protect the sides from future injury. When this operation was completed, the tooth presented very much the appearance of having had a close-fitting shell crown applied. Gold fillings were introduced in the palatal and approximal surfaces of the left cuspid, to protect it from the attachment. Gold crowns were placed upon the molars, one of which, the left, was pulpless. The appearance of the teeth after these operations is shown in Fig. 538.

Fig. 538.

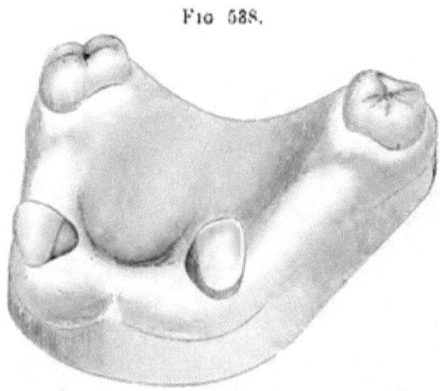

A plate bridge (Fig. 539) was then constructed, the attachments for which were collars on the molars, and half-collars or clasps on the cuspids. The former rested on shoulders formed on the gold crowns, and the latter on the palatal curves of the cuspids.

Fig. 539.

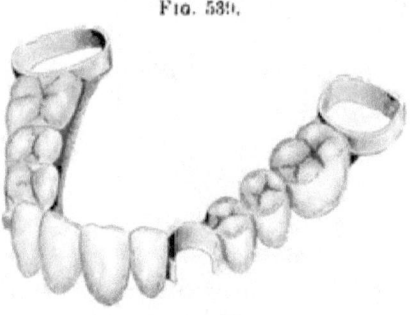

Fig. 540 illustrates a case as presented for treatment to Dr. H. A. Parr. The loss of the posterior teeth of the lower jaw and the abrasion of the anterior teeth had resulted in the abnormal occlusion shown. The incisive edges of the lower teeth were lengthened with gold contour fillings. Gold collar crowns with porcelain fronts were placed on the upper anterior teeth, to lengthen them sufficiently to restore the occlusion. The upper molars, which were all more or less decayed and broken down, were restored in form with gold crowns. The space

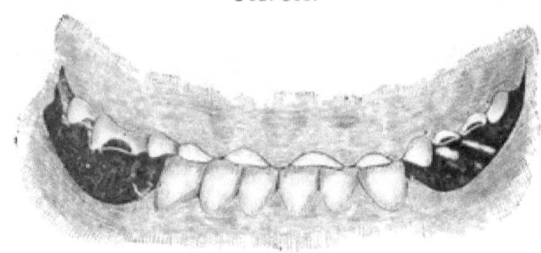

Fig. 540.

representing the loss of the left bicuspid was filled with a bridge tooth having a porcelain front, attached to the approximal gold crown. In the lower jaw a partial set was inserted on each side

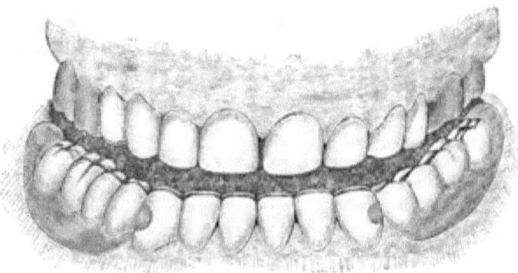

Fig. 541.

to substitute the lost posterior teeth. Fig. 541 shows the appearance of the case when completed.

The following case is a marked illustration of extensive contour filling combined with gold and porcelain crown-work. The operations were performed by Dr. E. P. Brown, with the exception of the porcelain inlays, which were inserted by Dr. C. H. Land. Fig. 542 represents the case before treatment. The teeth show the effects of erosion, abrasion, and decay. The upper teeth at the incisive and palatal portion were contoured with rolled gold, No. 60, condensed with the Bonwill electric mallet.

Fig. 542.

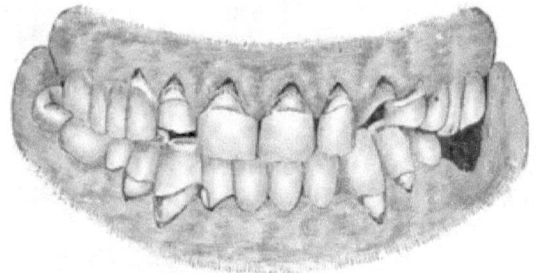

Fig. 543.

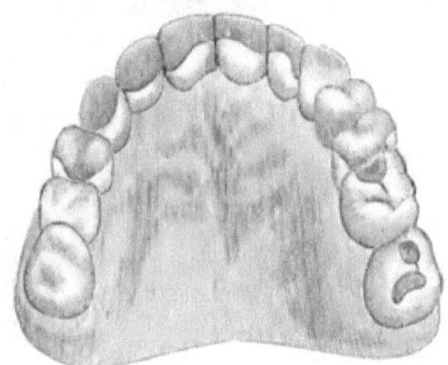

Fig. 544.

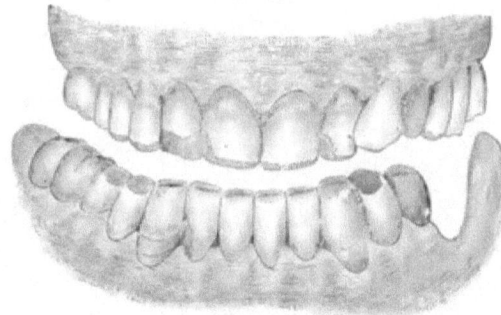

A porcelain crown was placed on the root of the left cuspid, and a gold crown on the first left bicuspid. The eroded, decayed, and abraded portions of the other bicuspids were contoured with gold. The treatment of the lower teeth consisted in contouring with gold the abraded occluding surfaces of the right first and second bicuspids and left first bicuspid. The left second bicuspid and two right molars were crowned with gold. The cervical decay which affected the inferior right lateral, cuspid, and first bicuspid and left cuspid was removed and inlaid with sections of porcelain colored with gum enamel at the line of the gum margin by Dr. Land. Fig. 543 shows the superior arch from the palatal aspect when completed. Fig. 544 is a labial or front view of the several operations.[1]

---

[1] The patient, Dr. W. I. Thayer, exhibited these operations, nearly completed, at the clinic of the First District Dental Society of New York, February, 1889.

PART IV.

MATERIALS AND PROCESSES USED IN CROWN- AND BRIDGE-WORK.

CHAPTER I.

PLATE AND SOLDERS.

Plate.—In the construction of crown- and bridge-work, gold, platinum, or iridio-platinum is used in the following forms and grades in carat, as suits the requirement of the case and the preference of the dentist:

Gold plate, 24 carats, from No. 31 to No. 34 U. S. standard gauge,[1] for constructing the collars and caps to collars of crowns and seamless crowns, and for backing porcelain fronts.

Gold plate, slightly alloyed,—about 23 carats fine,—in crown-work, if less flexibility of the metal is required.

Gold plate, 22 carats, No. 32 to No. 34 standard gauge, in constructing collars for crowns with porcelain fronts and all-gold crowns made in sections.

Gold plate, 20 carats, in construction of bridge-work.

Gold for constructing collars should be of as high a carat as possible, to better resist the action of acids. A large proportion of copper as an alloy is objectionable, owing to its tendency to cause tarnishing of the collar where an acid condition of the saliva exists. For this reason, U. S. gold coin, so much used for collars, is not quite suitable. It is also too stiff, and collars made of it are not readily burnished to fit at the edges. Dr. J. J. R. Patrick's formula, which is equal parts of coin and pure gold, affords a plate which is decidedly preferable, inasmuch as the proportion of copper is greatly reduced.

Gold plate, 24 carats, with a very thin lining of platinum or iridio-platinum plate, can be used in any of the processes connected with crown- and bridge-work, and is especially recommended to the inexperienced, for the reason that the melting of a part of a crown in the soldering process is prevented by its use.

---

[1] The United States standard is the only gauge referred to in the original descriptive matter of this treatise.

It is formed by placing together an annealed gold plate and a platinum plate (the gold about No. 20 gauge and the platinum No. 30) and passing them through a rolling-mill, in which process the plates are welded and reduced to the desired thickness.

Platinum and iridio-platinum plate is used for forming crowns for use in connection with porcelain bridge-work. Iridio-platinum plate for forming small collars need not be over No. 35 American gauge. At this thickness it can be easily adapted to the cervix of the tooth.

Pure platinum rolled very thin is used for forming caps to collars, backing teeth, and for various purposes connected with this class of operations. Iridio-platinum wire is used for pivots, pins, or posts, being more rigid than pure platinum. A wire of gold and platinum alloy is used on account of its elasticity for split or spring pivots or posts in removable bridge-work.

Solders.—Gold solder, 22 carats fine, is used for crown-work; 20 carats for crown- and bridge-work.

18 carats is used for crown- and bridge-work, but this is considered by some too low a carat.

14 carats for strengthening seamless crowns: used only in crowning operations.

Successive grades of solder from hard to easy flowing can be used in the regular soldering of crown- and bridge-work; gold plate or a hard-flowing solder for the first, a medium-flowing solder for the next, and an easy-flowing one for the finish. This avoids melting or flowing of the gold at any point previously soldered.

Any grade of solder can be made according to the following formula:

        Zinc, $1\frac{1}{2}$ grs.;
        Pure gold,
        Silver solder, in quantity sufficient to make up the $22\frac{1}{2}$ remaining parts in weight.

The quantity of silver solder used will regulate the grade in carat of the solder. Thus:

        Zinc, $1\frac{1}{2}$ grs.;
        Pure gold, 20 grs.;
        Silver solder, 3 grs.,

will, by the burning out of a portion of the zinc in the process, make a solder about 20 carats fine.

By lessening the proportion of zinc from 1½ grains to 1 grain, the proportion of silver solder being kept the same, the solder will become harder-flowing and a little finer.

Dr. W. H. Dorrance recommends the following formula as an alloy for the formation of different grades of gold solders, the proportion of the alloy used determining the melting point and fineness in carat of the solder:

>1 part pure **silver**;
>2 parts pure **zinc**;
>3 parts pure **copper**.

The silver and copper are first melted together in a crucible lined with borax and the zincs added quickly in small pieces, stirring the mass meantime with a pipe-clay stem. It is then, on the fumes of the zinc passing off, immediately poured into an ingot-mold or into a large wooden pail filled with water; 4 grains of this alloy melted with 20 grains of pure gold will result in a solder fully 20 carats fine.

As a solder for crown- and bridge-work constructed of 22-carat gold plate, Dr. Litch's formula is as follows:

>Gold coin (ten dollars), 258 grs.;
>Spelter (or brazier's solder), 24 grs.;
>Silver coin, 24 grs.

This is a proportion of about 26 grains of pure copper and 232 grains of pure gold to the remaining 48 grains of the alloy, and makes a gold solder a little over 18 carats fine.

Dr. C. M. Richmond originally used American gold coin for forming gold crowns, rolling it out in the form of plate. The scraps he formed into solder by melting and adding one-fifth of their weight of fine brass wire cut in small pieces, using plenty of borax.

Dr. Low's formula for solder in bridge-work is:

>1 dwt. **coin gold**;
>2 grs. **copper**;
>4 grs. **silver**.

This makes a solder about 19 carats fine.

The following formula[1] gives a 20-carat solder which is specially recommended for crown- and bridge-work:

> American gold coin (21.6 carats fine), $10 piece, 258 grs.;
> Spelter solder, 20.64 grs.

Prepared Solder Filings.—Prepared gold solder filings are made by filing with a clean, flat plate file a thick piece of solder held in a vise. The filings are allowed to fall in a box or on a sheet of paper placed to receive them. A magnet should be passed through the filings to remove any minute particles of steel. To five parts of the filings so made is added and well mixed with them one part of the prepared flux or finely pulverized vitrified borax. Solder prepared in this way is useful for strengthening crowns, and also in fine soldering operations, as the particles of the solder take the heat separately and fuse much more quickly than when the solder is cut in pieces. The flow of the solder is also more easily limited.

---

[1] American System of Dentistry, vol. iii, p. 849.

CHAPTER II.

PORCELAIN TEETH.

The qualities specially requisite in the body of porcelain teeth for use in crown- and bridge-work are density, strength, and the ability to withstand unaltered in form or shade any degree of heat to which they may necessarily be subjected. In these respects the porcelain teeth of our best American manufacture seem to excel, besides affording the most artistic imitation of the natural teeth in form and shade. They are also distinguished by the practical location of the pins.

In some crowning operations, where to imitate the conformation of a natural crown considerable alteration of the labial surface of a porcelain front is required, teeth of English manufacture may be used, as the texture of the porcelain admits of a fine polish being given to a ground surface.

Teeth are sometimes fractured in the process of soldering, caused by the contraction of the backing when adapted over the edges of the porcelain in a curve instead of at a right or slightly obtuse angle, or by melting solder on some point of the porcelain which is unprotected by a backing of metal. The solder, or the borax, as it cools, contracting on the porcelain, or a very thin edge of the metal covering it, will usually cause a fracture. The porcelain tooth has yet to be made that will, as a rule, endure such extreme treatment without breaking.

CHAPTER III.

MOLDS AND DIES.

METALLIC models of fusible metal can be easily and quickly formed for use in crown- and bridge-work. The melted alloy can be poured into a plaster, moldine, or gutta-percha impression taken in a tube or impression-tray. When a tube is used, a strip of paper should be wound around it to lengthen the die.

The following fusible alloys of tin are suitable for the purpose:

PROPORTIONS OF METALS.			MELTING POINT OF THE ALLOY.
Tin.	Lead.	Bismuth.	Fahr.
1	2	2	236°
5	3	3	202°
3	5	8	197°

Dr. G. W. Melotte, of Ithaca, N. Y., to whom is accorded the credit of introducing the use of fusible metal and the compound called "moldine" into crown- and bridge-work, gives the proportions of his alloy in parts as—

Tin, 5; Lead, 3; Bismuth, 8.

Dr. Melotte's moldine, a preparation compounded of potter's clay and glycerin (to which, when needed to soften it, more glycerin can be added), is very useful in molding.

A counter-die to a small cast or die of fusible metal is made by indenting a block of lead with a punch, and then driving the cast or die into it. Its use in crown-work is described on pages 96, 104, and 108, and by Dr. Melotte on page 233.

The following method of forming a metallic model of a prepared root or crown is given by Dr. W. C. Barrett, of Buffalo, N. Y., who accords Dr. H. A. Baker, of Boston, the credit of being the originator of it:

"Copper is rolled down quite thin, and a band three-fourths

of an inch wide wrapped about the root and forced up under the gum. A ligature is passed around both; the copper band is burnished down and the ligature drawn tight. The copper band will now fit just as we want the gold band to do. Plaster of Paris is then inserted in this, forced up against the end of the root, and permitted to set. Take it off, and if you use Babbitt-metal, a piece of paper wrapped about it (the copper band) will lengthen it out sufficiently, when the metal may be poured into it, and thus a perfect model of the end of the root will be secured. That part which is inserted in the copper tube is the exact reproduction of the root of the tooth. The model will perhaps need a little dressing down with a file, when the gold band may be fitted around it and soldered, thus avoiding the necessity for the annoying and painful trying-on in the mouth."

Fusible metal can be used instead of Babbitt-metal.

Fig. 545.

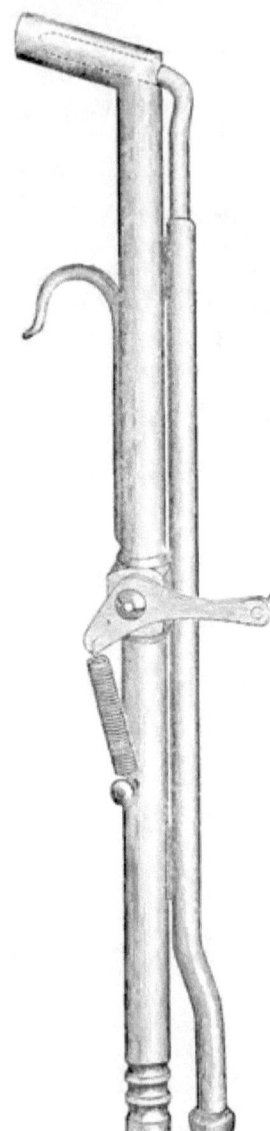

CHAPTER IV.

SOLDERING.

Investments for soldering, and models on which the soldering is to be done for bridge-work, are made in calcined marble-dust and plaster, in the proportion of equal parts for models, and one part of plaster to two of marble-dust for investing. Sulphate of potassium, used in place of common salt, but in smaller quantity, will cause the mixture to set quickly and hard.

Soldering.—In soldering crown- and bridge-work or sections of it containing porcelain fronts, the investment should be first uniformly and thoroughly heated, and the heat maintained during the process of soldering. This is especially necessary in the final soldering of gold crowns with porcelain fronts, as the solder is thereby flowed into the interstices between the porcelain and the caps and gives continuity of structure to the crown. A large piece of charcoal, concave on the side to be used, furnishes a good soldering support, as it retains the heat. A hand gas blowpipe, operated by a foot bellows, and so constructed that the flame is under perfect control, is the most suitable (Fig. 545).

The use of Borax.—In soldering invested sections of a crown or bridge together, borax which has been reduced to a vitreous state by heat and then finely pulverized is the preferred form for its use. Prepared in this manner it should be sprinkled on the part previous to the commencement of the soldering, and a little added at any time it is needed during the continuance of the process.[1]

In small, fine soldering of invested or uninvested work, the lump borax ground with water on a slab to a cream-like consistence and applied with the point of a stick or brush only where it is desired to have the solder flow, is the most suitable.

---

[1] Dr. Parr's "prepared flux," a finely pulverized vitrified flux, can be used in this manner. It is conveniently kept in the form of a saturated solution in a bottle, from which the operator can use it with a brush or stick in fine solderings.

It is also prepared in combination with wax cement. In the melting out of the wax when the case is invested and heated for soldering, the flux is carried into the interstices.

CHAPTER V.

INSTRUMENTS AND APPLIANCES.

The dentist who intends to engage extensively in crown- and bridge-work, and who desires to practice it conveniently and successfully, should supply himself with all the necessary instruments, tools, and other appliances. These consist principally of

Fig. 546.

drills of various styles, among them the Gates-Glidden for root-canals; root-trimming and shaping instruments; corundum-wheels and points; rubber and corundum disks; a set of steel mandrels, with a contracting plate and suitably shaped pliers; a supply of clamps of various forms for use in the many soldering processes; some moldine and fusible alloy, and an apparatus for forming gold caps for use in constructing crowns; a Lee blow-pipe or Knapp's compound blow-pipe, or both, for use as is most suitable or convenient.

The bench on which the principal parts of the work are done should be specially designed and reserved for it. It should be very convenient to the dental chair, and if in the same room

FIG. 547.

should be situated a little behind it, out of view of the patient. Everything connected with this bench should be kept in order and ready for immediate use. Such a bench, made of black walnut, with a top that can be closed when not in use, making an unobjectionable piece of furniture, is represented in Figs. 546 and 547.

ABRASION of incisive edges of teeth, 102
  method of forming crown for, 122
Abscess, chronic alveolar, treatment of, 35
Acid, arsenious, its use and action in devitalizing pulps, 28
Acid secretions, platinum unaffected by, 123
Adjuster for use in cementation of crowns, 143
Adjustment of finished bridge-work in the mouth, 160
Alloy for forming any grade of gold solder, 283
Alloys of tin, their use in crown-work, 286
Alveolar abscess, classification of, 35.
  treatment of, 35
Ames's method of forming gold tips, 136
Amputation of the apex of a root, 37
Analysis of dentine, 24
Anatomical structure of dentine, 23
Anchorage bars in bridge-work, 157, 174, 263, 281
  manner of attaching to the abutments, 153–159, 264
  manner of forming, 159, 175
Anchorages for bridge-work, preparation of, 153, 163, 281
  Dr. Litch's method, 236
  Dr. Parr's, 200
  shell, 166
Ancient bridge-work, 147
Anesthesia in pulp-extraction, 27
Anesthetics, local, for application to gum, 36–38, 88
  use of in crown-work, 88
Antagonizing teeth, preparation of their cusps, 42
Antiseptic agents in treatment of alveolar abscess, 37
  in treatment of pulpless teeth, 33
Articulation for bridge-work, manner of taking, 155
Artificial crown-work, 17, 45
  the gold system, 82
  the porcelain system, 48
Artificial gum in porcelain bridge-work, 267
Attachments for removable bridge-work, 208, 218

BACKINGS for porcelain fronts in crown-work, 90, 285
Baldwin's method of mounting crowns, 76
Bars for bridges, 165
Beers's crown, 84

Bing's bridge-work, 148
Blow-pipes, Lee's, 288
  Knapp's carbo-oxyhydrogen, 230
Bonwill's porcelain crowns, 49
Borax, method of using in crown- and bridge-work, in constructing root-caps and tubes, 207, 288
Bridge-work, 145
  adjustment and insertion, 160
  an impartial criticism of, 149, 150
  as affecting hygienic condition of the mouth, 187
  cantilever, 173
  cases illustrating the application of, 206
  cementation of, 143
  construction of, 152
  detachable, 189
  double-bar, 174
  extension, 170
  foundations for, 152
  manner of taking impression and articulation for, 155
  mechanical principles governing the process of construction, 153
  partial cap and pin, 236
  porcelain, 262
  removable, 189
  removal of, 186
  selection of abutments, 152
  with replaceable porcelain fronts, 175.
Brown's porcelain crowns, 69
  bridge-work, 262
Burnishers for adapting collars, 88

CAPPING pulps, methods of, 24
Cementation of crown- or bridge-work, 142
Chronic alveolar abscess, 35
Clamps, soldering, 96
Collar contractor, 248
  crowns hygienically considered, 122
  pliers, 249
Collars for crowns, 84
  construction and adaptation of, 84
  Townsend's fusible die in forming, 74
Construction of bridge-work, 152
  detachable and removable, 189
  mechanical principles governing, 152
  plate, 206.
  saddles, 170, 181
  single and double bar, 173, 174
  small cases of, 161
  special processes and appliances in, 163
Corundum-wheels and points, 39, 40

Crown- and bridge-work combined with operative dentistry in dental prosthesis, 272
  instruments and appliances, 290
Crowns, artificial:
  Baldwin's method of mounting, 76
  Bonwill, 49
  Bonwill cap, 77
  Brown, 69, 102
  Farrar's cantilever, 132
  Foster, 62
  Gates, 62
  How, 55
  Howland, 63
  Kirk's method of mounting, 75
  Leech, 125
  Logan, 64
  Low, 126
  New Richmond, 70
  Parr, 124
  Patrick, 102
  Perry, 128
  Richmond, 200
  Rynear, 103
  Stowell's method of mounting, 93
  Van Woert, 83
  Weston, 78
  all-gold, in sections, 95
  attachments for all-gold and seamless gold, 42, 43, 117, 118
  cementation, process of, 142
  contouring of collars, 112
  countersunk, 94
  dies for use in construction of, 97
  expanding, 114
  finishing and polishing, 142
  for abraded teeth, 122
  for separate molar roots, 132
  gold and porcelain, for teeth with living pulps, 119
  gold and porcelain, without a collar, 82
  gold collar, 84
  gold seamless cap, 104, 113
  gold seamless contour, 110
  mandrel system, 245
  partial, 134
  porcelain, with collar attachment, 74
  porcelain, with rubber attachment, 81
  preparation of crown or root for, 21, 39
  process of adjustment of gold contour, seamless, 113
  remarks on the use of collar or porcelain crowns, 72, 122
  removal of, 186
  repair of, 185
  shell, 166, 171
Cusps of antagonizing teeth, preparation of, 42

DETACHABLE bridge-work, 189
Detachable porcelain front, 175
Devitalization of pulps, 27
  heroic or instantaneous, 27
  use of arsenic for, 28
Die-plate, 97
Dies, 92, 95, 96, 97, 104, 107, 108, 120, 286
  counter, 96, 108
  Dr. Baker's method, 286
  Dr. Melotte's method for forming, 232
  fusible metal, 286
  manner of obtaining molds and dies, 104, 286
Diseased pulps, classification of, requiring extirpation, 24
Disks, forms of, 40
Double bar-bridges, 174, 265
Drills, Gates-Glidden, form of and method of using, 31
Dummies, definition of, 158
Dwinelle's crown, 84

EXCISION of crown, 27
  when to avoid, 41
Expansion of a collar or crown, 114
Extension bar-bridge, 267
Extension bridges, 170, 267
Extirpation of pulps, 24, 27

FARRAR'S cantilever crown, 132
Ferrules for root-crowns, 84
Files for trimming roots or crowns, 40
Filling of root-canals, 31
Finishing and polishing bridge-work, 159
  crown-work, 142
Forceps for excising crowns, 28
  for repairing, 185
Formulas for fusible metals, 286
  for gold solders, 282
Foster crown, 62
Foundations for bridge-work, 152
Fracture of porcelain teeth in soldering, 285
Fractured teeth and roots, treatment of for crowning, 130
Fusible alloys of tin, 286
  Melotte's, 286

GATES'S crown, 62.
Gold collar crowns, 84
  preparing natural teeth for, 39
Gold, method of using, 115, 121
Gold plate lined with platinum, 281
  crown-metal, 281
  solder fillings, 282
  solders, formulas for, 282
  standard of carat and gauge required, 281
  tips, Dr. Ames's method, 136
  wire, 282

INDEX

Gutta-percha for forming molds of crowns or roots, 104
  in filling root-canals, 33
  in preparation of roots, 42
  use of, for cementing crown- and bridge-work, 144

HEAT, use of as a disinfectant agent, 24, 32
Hollow wire for posts, 90
How crowns, 55
How screws, 55, 62
How's root-trimmers, 40
Howland crown, 63
Hub-mold, 97
Hygienic condition of the mouth as affected by bridge-work, 187
Hygienic consideration of collar crowns, 122
Hygienic preparation of the mouth, 24

IMPRESSIONS of crowns or roots, 51, 104, 107
  cups, 91
  for bridge-work, 155
  impression and articulation combined, 155
  materials for taking, 91, 155
Instantaneous devitalization, 27
Instruments and appliances, 290
Investments for soldering, 288
  in bridge-work, 158
  in crown-work, 91
Iodoform, methods of using, 33
Iridio-platinum wire for posts, 282
Irregularities of the teeth, methods of crowning in, 133

JUDICIAL decision regarding the public use of bridge-work, 222

KINGSLEY'S method of forming all-gold crowns, 97
Kirk's method, 75
Knapp's methods in crown- and bridge-work, 227

LAND'S method in partial porcelain crown-work, 139
Lead counter-dies, method of forming, 96, 105, 108, 286
Leech's crown, 125
Litch's method of crowning, 106
  detachable bridge, 192
  partial cap and pin bridge, 236
Logan crown, 64
  Dr. Baldwin's method of mounting, 76
Low bridge, 222
Low crown, 126

MANDREL system, 245
Mandrels for forming collars, 85, 247
Materials and processes used in crown- and bridge-work, 281
  molds and dies, 286
  porcelain teeth, 285
  soldering, 288
Melotte's method, 232
  metal, 286
  moldine, 286
Metallic dies and counter-dies, 286
Metallic dies for forming caps with cusps for crowns, 97
Models for bridge-work, 155
Moldine, 286
Molding, methods and materials used in, 95, 104, 107
Molds and dies, 286
Morrison's crown, 84
Mouth, preparation of, 24

NECKS of teeth, average forms of, 85
New Richmond crown, 70

OBJECTIONS urged against bridge-work, 149
  against collar crowns, 122
Ottolengui root-reamers and facers, 66
Oxyphosphate cement, 137

PARR'S crown, 124
Parr's detachable and removable bridge-work, 200
Partial crowns, gold, 134
  Dr. Ames's method, 136
  Dr. Land's method, 139
  Dr. Littig's method, 138
  Dr. Parr's method, 136
  porcelain, 137
Patrick's cap-stamping machine, 112
  crown-work, 102
  formula for gold for collars, 281
Perry's crown, 128
Pin-bridge, partial cap and, 236
Plaster impression and articulation, method of taking, 155
Plate and solders, 281
Platinum plate, 282
  unaffected by acid secretions, 123
  wire, 282
  with gold, 282
Pliers for shaping collars, 85
Porcelain and gold crown without a collar, 82
Porcelain bridge-work, 262
Porcelain crown with gold collar attachment, 74
Porcelain crown with rubber or vulcanite attachment, 81

Porcelain crowns, remarks on the use of, 72
Porcelain faces for crowns, 155, 175
Porcelain fronts, backing for, 90, 285
Porcelain teeth, selection of, 285
    some causes of fracture in soldering, 285
Pouring fusible alloy or metal, manner of, 95, 104
Preparation, special, of badly decayed teeth and roots, 42
Pulp, diseases of, requiring extirpation, 24
    instantaneous devitalization, with excision of crown, 27
    preservation or devitalization, 23
Pulpless teeth, their treatment and disinfection, 31
Punch forceps for riveting, for use in repairing bridge-work, 185

REMOVABLE or detachable bridge-work, 189
    Dr. Litch's, 192
    Dr. Parr's, 200
    Dr. Richmond's, 200
    Dr. Starr's, 194
    Dr. Waters's, 218
    Dr. Winder's, 189
    methods of forming attachments for, 202
Removable plate bridges, 206
    Parr's methods, 216
    Waters's methods, 218
Removal of crown- and bridge-work, 186
Repair of crown- or bridge-work, 185
Retaining-pin for all-gold crown, 117, 118
Richmond crown, 89
Root-canal, antiseptic agents for treatment of, 33
    dryer, 32
    method of filling, 32, 33
    method of treatment, 31
Root-reamers and facers, Ottolengui's, 66
Root-trimmers, Starr's, 197
Roots intervening between abutments, 166
Roots, special preparation of badly decayed, 42
Rubber or vulcanite attachment for crown, 81
Rules governing the insertion of bridge-work, 152
Rynear's crown, 103

SEAMLESS gold collars, 246
Seamless gold crowns, 104
    method of contouring, 112
    method of forming from an impression, 107

Seamless gold crowns, process of adjustment and insertion, 113
Screws for use in crown-work, 55, 62
Sectional crowns, 89
Sections of bridge-work, construction of, 159, 162, 164, 192
Self-cleansing spaces, 156
Shaping teeth and roots for crowning, process of, 39
Shell anchorage or crowns, 166
    seamless, 168, 240
Shoulders on the anterior teeth, 163
Slots for anchorage bars, 154, 164
Solder, gold, formulas for, 282
    Dr. Dorrance's, 283
    Dr. Litch's, 283
    Dr. Low's, 283
    Dr. Richmond's, 283
Soldering, manner of, 288
    investments for, 288
Solid gold crowns, 163
Special forms and methods in crown- and bridge-work, 124
    Dr. Knapp's, 227
    Dr. Low's, 126, 222
    Dr. Melotte's, 232
    mandrel system, 246
Special preparation of badly decayed teeth or roots, 42
Spur support in bridge-work, 172
Stamping press for caps, 112
Starr's method of detachable bridge-work, 194
    root-trimmers, 197
Strengthening gold seamless crowns, 116
Syringes, hot-air, 32
    abscess, 36

TABLE of fusible alloys, 286
Teeth, porcelain, 285
Temporary attachment of bridge-work, 144
Thickness of plate suitable in crown-work, 85
Tin, alloys of, 286
Townsend's fusible die for forming collars for porcelain crowns, 74
Treatment of chronic alveolar abscess, 35
    preparatory, of the mouth, 130
Trying in bridge-work, 156

WARPING of bridge-work in soldering, 160
Waters's removable bridge-work, 218
Weston's crown, 78
Winder's detachable bridge-work, 189
    Dr. Sharp's method in, 192
Wire for posts or pivots, 90, 282

ZINC, oxyphosphate of, 142

www.ingramcontent.com/pod-product-compliance
Lightning Source LLC
Chambersburg PA
CBHW032052220426
43664CB00008B/972